TRAPPED BY TOURISM

TRAPPED BY TOURISM

Sustainability Questions for a World Fueled by Travelers

LARRY KROTZ

ROWMAN & LITTLEFIELD
Lanham • Boulder • New York • London

Published by Rowman & Littlefield
An imprint of The Rowman & Littlefield Publishing Group, Inc.
4501 Forbes Boulevard, Suite 200, Lanham, Maryland 20706
www.rowman.com

86-90 Paul Street, London EC2A 4NE

British Library Cataloguing in Publication Information available

Library of Congress Cataloging-in-Publication Data
Names: Krotz, Larry, 1948- author.
Title: Trapped by tourism : sustainability questions for a world fueled by travelers / Larry Krotz.
Description: Lanham, Maryland : Rowman & Littlefield, 2024. | Includes bibliographical references and index.
Identifiers: LCCN 2024027491 (print) | LCCN 2024027492 (ebook) | ISBN 9781538196465 (cloth) | ISBN 9781538196472 (ebook)
Subjects: LCSH: Tourism--Environmental aspects. | Sustainable tourism.
Classification: LCC G156.5.E58 K76 2024 (print) | LCC G156.5.E58 (ebook) | DDC 338.4/791--dc23/eng/20240701
LC record available at https://lccn.loc.gov/2024027491
LC ebook record available at https://lccn.loc.gov/2024027492

CONTENTS

Acknowledgements vii

Introduction: Is One of the World's Biggest Industries Sustainable? . 1

CHAPTER 1: No Surprises: Why We Travel 11

CHAPTER 2: Cuba: The Idiosyncratic Island That Bet
Big on Tourism. 25

CHAPTER 3: What Happens When Too Much Is Too Much?:
Over-tourism in Venice, Rome, Barcelona, and Granada 45

CHAPTER 4: Cruise Ships: Blessing or Curse?. 63

CHAPTER 5: English Cathedrals: Churches, Museums, or
Amusement Parks?. 77

CHAPTER 6: Winery Tourism: What It Really Costs 91

CHAPTER 7: What Does Sustainable Tourism Look Like? 105

CHAPTER 8: Being Maasai—and Engaging Tourism 117

CHAPTER 9: Indigenous Tourism: Can Indigenous
Communities Get in on the Action and Still Preserve Their
Cultures and Lands?. 127

CHAPTER 10: What Digital Photography Has Done to Tourism . . 143

CHAPTER 11: Tourism Futures. 151

Notes. 163

Bibliography 171
Index 173
About The Author 177

Acknowledgments

This book found two champions, first in my agent at Transatlantic, Brenna English-Loeb, who believed in and supported it from the get-go, and then in Rowman & Littlefield managing editor Deni Remsberg, who quickly scooped it up for publication. My heartfelt appreciation to both of them.

While the work of writing was going on, I received support from numerous friends and associates. Donald Walker sent regular alerts and updates about the world of global tourism; David Berlin read an early manuscript, providing useful feedback. Beyond that, Stephanie, John, Hilda, Christopher, Alain, Geoff, Debra, Kevin, Sherali, Naomi, Neil, Vince, and many others all shared generously of their perspectives and experiences. I dedicate this final product to all of them.

Is One of the World's Biggest Industries Sustainable?

IN THE EARLY DAYS OF MARCH 2020, MY WIFE AND I WERE NEARING the end of a ten-day stay in Cuba. It had been delightful: time on the beach, cultural entertainments in the evenings, a day-long bus tour across the island. But by day eight, CNN and the Internet that was available were delivering disturbing news about a rapidly emerging health threat, enough information that vacationers—us among them—started to pack up our sunny all-inclusive resorts in a panic. We boarded an already crammed bus and, as it lurched its way to the airport—which was a total frenzy of people cutting holidays short and fleeing to the presumed safety of "back home"—I will never forget how our tour guide broke into sobs. Juan knew he held one of the most cherished jobs in his country, escorting groups of foreigner visitors. Would he, in a million years, ever get it back?

The COVID pandemic starting at that moment in 2020 brought an abrupt halt to many aspects of "normal" life—one of them the world of travel and tourism. By so doing, it alerted us to a couple significant issues, neither of which we had until then very much considered. Though at almost opposite ends of the spectrum, these are incredibly important each in their own way.

The first is that as things shut down and people—like Juan—started to lose their jobs, we noted the substantial place tourism held in not just our domestic but the global economy. Airlines, hotels, restaurants, bus and limo companies, cruise lines, and sightseeing guides together made up substantial parts of our economies, and when that all shut down, it created a massive hole.

At the other end, as the pandemic settled in we began seeing reports about how the air—all over the world—was suddenly cleaner. We learned how with the drastic reduction in airline flights, the amounts of greenhouse gasses were lowering; with fewer cruise ships, the seas were cleaner. Whales and sea lions were happier. Cities like Venice were stunned to note that they were now quiet and clean. Diminished noise and congestion made Venetians better able to enjoy their own city.

Stark realizations on one side, silver linings on the other. And a substantial tension in between. The pandemic showed how dependent individual as well as the global economies are on the multibillion-dollar businesses of tourism and travel—quite possibly at unsustainable levels. But we were also better able to quantify the impact of tourism and travel on both the natural and cultural environments of the world—quite possibly also unsustainable.

Yet what are we to do? Travel—spending almost any holiday time we get going some place far away to have a look around or enjoy ourselves on some adventure—is what masses of us do consistently. Many of us come from cultures where not very many generations ago ordinary people weren't likely to venture more than thirty miles from the villages of their birth (unless they were in the army and got sent to some far-off war). My great-grandparents made one big trip in their lifetime: from their village in southern Germany to emigrate to America, a journey that consumed months. For the rest of their lives, the only major trip they made was to go by train from the small southern Ontario town where they had settled to visit relatives in Buffalo, New York. A mere three generations later, we think nothing of picking a spot on the opposite side of the earth and just dropping in for a couple of weeks.

Not just Canadians and Americans but Europeans, Asians, Latin Americans, certainly Australians, millions of us use our expendable income and our free time to visit one another's countries and places of interest or do exotic things far from home. According to the UN World Tourism Organization (UNWTO), more than two billion of us did this in the most recent year for which they have numbers, 2020, and paid out more than a trillion dollars to make it happen.

When I wrote about tourism twenty-five years ago, in 1996, I observed that two things made mass tourism possible: one was cheap air travel, the other was the credit card. That was just the beginning. As those intervening twenty-five years have come and gone, we have added all kinds of new technologies, such as vastly expanded possibilities for things like direct bookings via the internet where people make their own Airbnb arrangements. We came up with mobile phones and roaming plans that make it infinitely easier to work and holiday simultaneously or simply remain in very comfortable contact with home even when we might be far away. Our banking systems and currency rules streamlined the ways and amounts of money we can move around as we travel from country to country. We have come a long way—and very quickly—from the need for either travel agents or traveler's checks. Even the manners in which we learn about and make our decisions got finessed. The standard old Fodor's or Lonely Planet guidebook has been significantly supplemented by "influencers" who use online social media platforms to tell vast numbers of targeted people where they should be going and what they should be doing once they get there.

The place tourism holds in the economies of the world is a tricky one to quantify with exactitude. As an industry, travel and tourism includes hospitality of all sorts for people moving quite locally as well as those traveling halfway across the world. What is measurable and identifiable, though, are trends and the differences between what is going on now in comparison to what the case might have been a decade ago or a quarter century ago. One reliable gatherer of data is the World Bank, which compiles stats from other bodies such as the World Tourism Organization and whose *Yearbook of Tourism Statistics* is a compendium of statistics and data files. According to WTO figures, tourism receipts across the globe grew from $522 billion USD to $1.8 trillion between 1995 and 2020. A fourfold increase in money made. International tourism (this excludes travel within one's own country) clocked departures of 930,066,023 in 1995, rising to 2.03 billion in 2020. More than a doubling of the number of travelers.

With the help of a summing-up from tourismteacher.com, we can assert that the tourism economy represents 5 percent of the world's gross domestic product (GDP). It provides for 6 to 7 percent of total global employment. After fuels, chemicals, and automotive parts, tourism is the fourth-most-important product of international trade. The global tourism industry is valued at $1 trillion a year. In over 150 countries, tourism is one of the top export-earners and is the main source of foreign exchange for one-third of developing countries and half of less economically developed countries.

This is the world that became frozen when the COVID pandemic struck—and that people and governments in turn were desperate to rebuild once the pandemic ended. The numbers represented people's jobs and livelihoods the world over. They also represented the quantifiable output of the lifestyles and discretionary spending of many millions of us who are, for lack of a better term, tourism's consumers. And, if we go back to the World Bank comparisons made between 1995 and 2020, we can see that this world has been voracious about its growth, with pressures from almost every quarter to continue a never-ending expansion. These are the balls that are in the air and that need to be kept in the air when we talk about "sustaining" the tourism industry and, with it, constant travel as our way of life.

Sustainable, however, is a multifaceted concept.

* * *

In October of 2022, people locally in Vancouver watched on television as Canada's Minister of Tourism held a press conference on a pier in the city's harbor. It was the standard kind of photo op in which government officials announce the distribution of a dose of public money into the hands of some worthy enterprise. In this case, the enterprise was British Columbia's once-massive tourism industry now trying to recover from the COVID pandemic. And the $1.2 million CAD (a rather modest sum as such things go) was to benefit a handful of enterprises such as building a digital platform to be used to promote "adventure tourism" along with expanding a seaplane terminal. It all went smoothly until, while the cameras from local TV stations rolled, an unscheduled interloper jumped into the mix: a white-haired man charged forward to grab the micro-phone from those who were supposed to be questioning the minister and commenced screaming at the gathered dignitaries. Eighty-six-year-old David Suzuki, longtime environmental activist, was on his favorite topic: climate change. How, he shouted, gesticulating to the mountains in the background where puffs of white smoke from a summer of forest fires still lingered, could they possibly think about expanding tourism while all of this was going on? The natural beauty that was supposed to attract visitors was being destroyed before their very eyes.[1]

This book is about tensions. Tensions between what we want and what we do. Tensions between what is inevitable and what we can sur-vive. It will look at the issues not so much of "sustainable tourism" as "tourism's sustainability," both as it plays out in our economies and as a factor impacting our natural and cultural environments. We will never shut tourism down; we are destined to have it and to participate in it. But what truly are its implications for the world we live in? By delving into a number of examples we will look at how what we do and how we do it affects a handful of important corners of our world, and how awareness has developed about steering the impacts in ways that work for every-body. If communities and governments seek economic benefits, they must also look at the trade-offs they might be pressured into on the way to getting those benefits: commodification of cultures, economic unfairness, environmental stresses.

An intriguing tension is one we can find in the Caribbean island of Cuba. Due to its political isolation and strict ideology, Cuba might not seem a natural for allowing foreign tourists in great numbers, and yet has done so deliberately. The islands of the Caribbean are considered a paradise by northerners wanting sand and sun and a bit of exotica away from the cold of winter. Their coasts are littered with all-inclusive tourist resorts and cruise ship ports. But among them, the still sternly Commu-nist nation of Cuba remains uniquely intriguing. Substantially isolated by the strict economic and political embargo placed on it more than sixty years ago by its powerful neighbor, the United States, Cuba took a gamble more than three decades ago that one of the few routes open to inject foreign exchange into its economy might be tourism from those parts of the world not controlled by the US. Such places included not only Russia and China, but Canada, Mexico, Germany, and Italy. Cuba got busy building a system within which to do this, and that is what we will look at.

A second field of tension is the one that is growing rapidly in more than a few of the most iconic tourism destinations on the globe: the beautiful cities of Europe, particularly the old and accessible ones like Rome, Venice, and Paris. In those locales, beautiful architecture, splendid works of art, and cultural styles have been irresistible magnets to generation after generation of visitors who were born and raised in places perhaps less blessed. Those cities are what we have been told we need to put on our "bucket lists" and go to visit as soon as we are given the chance. Thousands upon thousands of us do so every year by whatever means we can take to get there: boarding airplanes as individuals or as groups on guided tours; by train or bus or cruise ship. The downsides, however (and the tensions), are to be seen in rapidly developing and sometimes dramatic signals that our tourism has outgrown its welcome. A great and growing number of the locals, apparently, don't want us anymore. They have had enough.

How are they doing this? What will it ultimately mean? One of these cities, Venice, "Jewel of the Adriatic," a couple of years ago took the dramatic step of halting cruise ships from docking in its lagoon. As a main port in the Adriatic, it was natural that Venice should become one of the major destinations for Mediterranean cruise ships when that form of touring became popular starting in the 1980s. When cruise ships became bigger, cruising fleets became larger, and the passenger lists became, shall we say, "less elite," Venice was one of the first places to feel itself overwhelmed. The ships came in and docked, the day-trippers spilled out, and the city reeled. Venetians themselves sensed that their city was being stolen from them and that it was becoming less and less affordable to live there. They began to question whether it was all worth it. The decision taken by the managers of Venice to put the brakes on all this is fascinating to study. How did it come to be? How does it work? Has it resulted in an equitable trade-off between tourism revenue and heritage conservation? How have other jurisdictions been influenced to try similar control measures where they live?

We will find another—and different—tension in Britain. There are forty-two Church of England cathedrals in England and Wales and twenty-four Roman Catholic cathedrals. Some Norman, some Gothic, built largely between AD 1000 and 1500, they are, without exception, massive, gorgeous buildings. Each is the seat of a bishop and a diocese, and each has its fundamental place in the history of both local and national England. Each has its story, but each likewise its modern-day financial challenges to stay afloat. These are no small matter. In each instance, what were constructed as churches and places of worship and ecclesiastical activity—and continue to want to remain that—have had to shift, for their continued survival, to more-unholy sources of income. Gloucester Cathedral in the west was able to rake in a great deal of cash as one of the sets for the Harry Potter movies on and off over a period of twenty years. Others have tried other things, not the least of which is some level of paid tourism. Durham, which was identified as a UNESCO world heritage site in 1986, had 727,367 visitors come through its turnstiles in 2019. Westminster Abbey in London (not a cathedral, though it once was) had 1.6 million visitors in 2019, each paying an entry fee of approximately 27 pounds. Tourists rather than worshippers have become the mainstay of the financial well-being and, in some cases, the very survival of a number of these splendid old cathedrals.

But what does that mean? What does it mean in terms of sustainability, including cultural sustainability? As in Venice, Rome, and Barcelona, is there, for the cathedrals of Britain, a point where there will be one tourist too many, a moment when the balance tips and what thinks of itself as still an active place of worship becomes purely a museum—or a theme park? Who is keeping an eye on that? Who is worried about it? And, by the same token, is anybody worried that no matter how many tourist visitors you might have and how high the entry fees, it will never be enough to keep the lights on and the roofs repaired in these very demanding structures?

For a fourth, and fascinating, tension we will go to the countryside. There the strains, growing in importance, coalesce around land use, particularly use of land that is considered precious. The country of Canada, my home, is blessed with one of the great natural wonders of the world. But unlike other major waterfalls such as Angel Falls in Venezuela or Victoria Falls in Zimbabwe, Niagara is right smack in the middle of one of the most heavily populated parts of not just the country, but the North American continent. For 150 years Niagara Falls has been a prime tourism destination, a must-see for both foreigners and the hundreds of thousands of Canadians and Americans who live within a two-hour drive of the spectacular cascade. And as a tourist site it sports all the accompanying paraphernalia to amuse the visitor and draw cash from their pockets: wax museums, boat rides, carnival-like arcades, casinos, theaters, revolving restaurants. The data for visitor numbers and economic impact is tricky to settle on for Niagara because it comes from two sources, Ontario Parks on the Canadian side and New York State on the US side. But generally agreed-upon numbers settle around thirteen million annual visitors and $3 billion in economic activity. A massive tourism enterprise.

Yet what is interesting about Niagara is that it is not the waterfalls and the Niagara River alone that have been turned over to the economy of tourism. For miles and miles, going back almost as far as the city of Hamilton (fifty miles away) and spanning the acreage between the Lake Ontario shore and the lift of the cliffs of something called the Niagara Escarpment, are scads of vineyards and wineries. Dozens of them. These are a relatively new feature, something largely of the last thirty years. Their purpose is important, but also what they replace is important. For more than a century, going back pretty much to early European settlement, the fertile belt of land, sheltered by the escarpment and warmed by the presence of the lake, has been prime fruit and vegetable growing territory. For a century and a half, Niagara was the produce section for Ontario's grocery stores: peaches, plums, pears, grapes. The area's orchards supported canning plants and jam and jelly factories. It was an invaluable garden supplying the tables of thousands of homes.

This changed drastically in the 1990s under the influence of two things: a North American trade agreement called NAFTA, which changed the continental rules around the production and trade of wine, and the growing influence of tourism as an economic driver. These two interventions led to a massive changeover in the region: orchards were torn up; wine grapes were planted. This is what forms our current history with two significant results. Yes, Ontario does produce some very nice wine, but the raison d'être for the plethora of wineries that have sprung up is not the wine they produce but the tourism trade they entice: hundreds of wine-tasting bars, winery tours, and wedding and conference venues. The economic impact of all of these is seen not under agriculture but under tourism. The question needing to be asked is, was this good? Is this sustainable? Households in Toronto now get their peaches from California and Mexico instead of from orchards close to home. Is that—including all the pollution long-distance transport entails—a desirable outcome? In Canada's vineyard regions, as in Venice, Britain, or the Caribbean, does tourism come with a hidden price tag? Is it one we are willing to take a look at? Is it one we are willing to reconsider paying?

Lastly, across the world, from the Maasai in Kenya and Tanzania to Indian nations in the United States and First Nations in Canada, Indigenous peoples have been turning to tourism in the hopes it might provide a route to economic self-sufficiency in ways preferable to building up heavy industry with its accompanying environmental costs. This is appealing to peoples who face, simultaneously, material poverty and pressures on their historic lands and cultural values. The Maasai, for example, have long been trying to negotiate accommodations that will allow them to remain living in those countries' national parks—and benefit from the safari tourism in those parks—while North American Indigenous nations have been hoping to strike a balance between economic benefit and protection of their cultural authenticity when *they* welcome tourists onto *their* lands. Such undertakings, at their best, can provide employment while allowing peoples to—under their own control—present their cultures to interested visitors.

The thing to watch is, as always, at what price? And how to keep things under control. In a number of places, people feel they haven't been able to do this. Many native Hawaiians, for example, feel that on their highly visited islands things have gone beyond both their control and their benefit. Everywhere, for Indigenous peoples who choose to go down this road, the watchword is to remain in control of both who a people is and what they are selling. To what degree should self-presentation be edited to conform to what is believed tourists "want"? In Kenya and Tanzania, apparently, young Maasai men and women are encouraged to hide their watches, blue jeans, and cell phones, as wearing them or carrying them would contradict the tourist expectation of the lone warrior with a spear, or the bare-breasted young woman bedecked with beads sitting in front of her hut. This is a bargain they have been pressured into. Along with this, to what degree are Indigenous peoples selling, in fact, their own marginality? All the issues of cultural commodification, economic fairness, and environmental pressures come into play.

Tourism has exploded massively since Thomas Cook, in the 1840s, took working people on weekend package train rides to the sea and also since Club Med, in the 1980s, invented the hermetically sealed all-inclusive experience. It has also exploded since the 1970s when anthropologists first started to think of travel and tourism as a human activity worth observing. Again, the continuation and growth of mass tourism is inevitable—there is no real turning back. But if government-promoted tourism is deemed the only way to keep the roofs on the cathedrals of Britain, or the economy of Cuba alive, or to provide Indigenous nations some economic independence, or to justify the maintenance of green spaces in Costa Rica, then that is a big issue to think and talk about.

Historically we used to think about travel as something that changed *us*. What it has become now, however, is something through which *we* change the world. This is a dramatic turnabout indeed. And it puts all kinds of novel questions and challenges on the table in front of us. What are our responsibilities when we travel? Are there things we should know about our destinations that we hadn't thought about before? What does being a responsible traveler or responsible tourist really imply? This book will try to be provocative on all those scores by looking at what we do through a variety of lenses.

Chapter 1

No Surprises

Why We Travel

I ONCE SAW A COMMERCIAL FOR A PACKAGE TOUR THAT ADVERTISED "NO surprises." This, I thought with a twinge of alarm, could not be right: Surely no surprises would be the antithesis of travel. Travel for me, up to that moment, had always been pretty much nothing but surprises. Many of them challenges to cope with—I often explain being on a trip as predominantly a series of problems to solve. You have to book places to stay, you search out transportation and make sure you get to the proper station on time, you lose your passport or your credit cards, you come down with a strange illness. I once was forced to rebook an entire trip to Africa because the train I was on, heading from London Victoria to Gatwick, got stalled for hours due to a reported bomb somewhere on the tracks. By the time I reached the airport, my flight to Harare had left and there wouldn't be another for three days.

Other surprises, of course, prove delightful. People you could never have imagined sit next to you on buses or trains. Hospitality you could not have anticipated lands in your lap. One time in British Columbia, a Kwakiutl family (granted I was there researching for a book and needed to interview them), instead of simply giving me an interview, invited me to join them on a visit to a modern-day version of a traditional West Coast potlatch. The best part of it: this was not nearby. Just bring my sleeping bag and a toothbrush, they said, because for the better part of a week the Billy family, Dan and Alberta, hosted me on their forty-foot fishing boat while we meandered along the glorious British Columbia coast from Campbell River north and then down a long channel into a remote community called Kingcome Inlet. Here, in a place accessible only by water or air, we ate plenteously of seafood, mingled with elders and their ceremonies, and enjoyed a three-day tournament of soccer matches.

Travel, both the glorious and the more challenging, is not—nor should it be—a set piece of predictability. Yet it is leaning more and more in precisely that direction. If we look at many of the ads we are confronted with, especially for things like cruises or all-inclusive resorts, predictability is what we are assured we can look forward to. This is meant to be reassuring. Tours, points out Dean MacCannell in his book *The Ethics of Sightseeing*, contain a jarring contradiction: "They expose tourists to new experiences and are designed to minimize risky fatefulness. The ultimate aim of every tour is to return home safely."[1] No surprises, therefore, is desirable for folks who might need a bit of coaxing to leave home in the first place. According to the ads, on ocean cruises everybody is either winning money in the onboard casino or posing with Mickey Mouse. On trips down some of the world's iconic rivers, the customers are all stretched out on lounge chairs with a drink in hand enjoying the view. "Nothing to worry about," claim the promos, "except to soak in the moments." What are the moments? Scenes of castles and vineyards wafting by as if one is watching a movie. The traveler is the passive receptacle of not so much experience as sensation. An entertainment.

Even this will potentially have new work done to it. Options opening in the future will expand to even more safe passivity. If we accept claims promised by Mark Zuckerberg, who previously changed the world by bringing us Facebook, all the adventure, exotic vistas, and wondrous sensations, in the not too distant future, could be achieved without ever having to leave home at all. In that new world we will inhabit the "Metaverse" wherein we will be able to enjoy not simply what our corporal lives can afford us, but additional activities on all kinds of virtual levels. While our flesh-and-bone bodies remain ensconced in our armchairs, or propped up in bed still in our pajamas, avatars representing us will be out attending concerts, marching through the world's most famous art galleries, hang gliding or parachuting through scenic vistas.

A great deal of what will be on offer will be facsimiles of what we now think of as travel. The promise is to take us virtually—but with such intensity that it will feel very "real"—across the savannah to mingle with a herd of elephants; right into the stadium to watch Real Madrid play a football match; jostling through a crowd in St. Peter's Basilica to have a look at David and the Sistine Chapel ceiling; scuba diving off the Great Barrier reef; or climbing Mount Everest (without any risk of adding to the 310 who have perished attempting the feat). Such prospects will prove truly unnerving for travel as it is configured today as an "industry" (nobody will need a hotel). It will be light years away from Rick Steves on PBS taking us on a guided tour through Switzerland. Yet it is both possible and to a high degree probable that some version of it will come to pass.

Much as some find this exciting to contemplate, might it not also be more than slightly weird? If it catches on big-time, life in the Metaverse will take huge numbers of us down rabbit holes not dissimilar to spending our waking hours in dark rooms watching soap operas. Rather than experience, this will be non-experience. It's hard not to see it like some aberrant form of porn.

Traditional definitions of "travel" are antithetical to what is essentially voyeurism as offered by "no surprises" or the Metaverse. The foundation of traditional travel is engagement—on multiple levels. The challenges are to be ready for experience while not knowing exactly what that experience is going to be. Classic travel's philosophy has been formulated over a number of centuries, predominantly through the writings of those who were both privileged enough and motivated to move around the world and keep track of their observations. These observations were equally of what was around them and what was going on inside their heads. The tenets of this philosophy embraced curiosity, engagement, informed education, and humility. Travel writers I've enjoyed have all been philosophers as well as travelers, explorers as well as reporters. Sir Patrick Leigh Fermor, Paul Theroux, Ryszard Kapuściński, Mary Kingsley, Bruce Chatwin. The travel of these—and indeed many young people trekking off to Asia in our own generation—is of discovery of both the world and of self.

Fermor, who died in 2011, was identified posthumously as "writer, scholar, soldier, polyglot." In *A Time of Gifts* (on foot to Constantinople, from the Hook of Holland to the middle Danube, 1977) and *Between the Woods and the Water* (1986), he reflected splendidly on travels he undertook as a young man through the Europe that was then on edge between the wars, jarred by the first and veering toward the second. It was 1933, and well connected through family and friends, the then-eighteen-year-old Fermor was able to get invited to dinner—and be given a bed—in some high-placed aristocratic home on one night while, on the next, needing to bed down in a haystack he might happen upon by the roadside. Oh to be young and full of bravado, but likewise open and embracing of whatever the day presented.

Charles Darwin, Alexander von Humboldt, and Mary Kingsley were explorers who set out, each of them, with some notion of what it was they wanted to look for, but also with the patience required for them to find it. Great tedium was part of the program that just had to be taken for granted. Von Humboldt, the German founder of academic biology as we know it, waited for weeks for the various ships that took him eventually to the jungles of South America, never certain that any of these were ever going to show up or set sail. In the meantime he had to both amuse and support himself.

Mary Kingsley set off in 1893 and 1895 to West Africa to collect zoological specimens and investigate the beliefs and customs of the tribes. As with von Humboldt, almost everything possible happened to her before she finally arrived at the "specimens." Earlier, in 1876, bad weather had reportedly kept her from being an observer of Custer's army getting ambushed by the Lakota Sioux at Little Big Horn in America's Montana Territory. When she got to West Africa, Kingsley got the locals to teach her survival skills before she headed into the wilderness. Like other explorers such as Darwin and von Humboldt, *intrepid* was the applicable word for her and her ventures. She died of typhoid at age thirty-eight while working as a nurse in a field hospital for Boer prisoners of war in South Africa.

Other travel writers I admire because of their attitudes as much as or possibly more so than the terrains they actually explore are Ryszard Kapuściński and Paul Theroux. Theroux quipped about the continent of Africa in *Dark Star Safari*, "all the news out of Africa is bad, it made me want to go there."[2] For those two the best strategy, and the one they promote to all of us, is to set off with expectations of nothing, but be adaptable and ready to fill in the blanks.

Our mental attitudes around travel are subject to a number of pitfalls. An ever-present danger are arrogances that can creep in and prove poisonous. An example of these is the urge to compare the place we are traveling through with what we have at home. You've possibly encountered this if not in yourself then in the chatter of fellow travelers. Unbelievably, nothing matches up with what they have left behind at home. However, if we are going to go to Cuba only to gripe that the internet is not as fast or accessible as what we have at home, or that the food is different or "not as good," why did we go there in the first place? This kind of kvetching is a true arrogance and destroyer of engagement.

Another arrogance is to think of ourselves as somehow above the rest of the tourist mob we have around us. This is a tough one to wrestle down, for I myself know the feeling of not wanting to be part of, and not wanting to be confused with, "the herd." From time to time I admit to nurturing the desire to pass myself off as or be mistaken for a local, to walk the other way when I see a tour group of fellow Canadians or Americans coming up the street or village road. I recognize the panicked feeling of not wanting to be caught in the restaurant crush and be mistaken for one of *them*. Once in Oxford, England, where my children were living for a period of months and I went to visit, my eight-year-old grandson and I were on a walk around town. In front of Christ Church Cathedral, we suddenly found ourselves trapped amongst a busload of tourists newly arrived from Korea. Spilling out of their bus, they were scrambling to take photos. Since I'd been there a few days and my grandson a few months, my instantaneous impulse was to determine that I could be properly annoyed by the picture takers because I was certainly not one of them. Wasn't I properly different and accustomed to the surroundings in so many ways? Yet to fancy myself more of a local was absurd—I ought to have realized that I was far more like a Korean tourist than I was an Oxfordian.

Perhaps the problem is that since the travel experience is so special, we correspondingly want to think of ourselves as special. That is, too special to be who we really are, especially since what we are usually is a confused visitor clutching a phrasebook or checking a phone to find a map. We indulge an extended fantasy, insisting that we are not like other people. But since we are not doing what we normally would do every day, neither are we quite ourselves. The reason this is confusing is because it is the opposite of what we ought to be doing. Travel should not be the way for us to forget or deny who we are; it should, rather, be a means of reminding us who we are: a foreigner, often unilingual and frequently insecure. That is its beauty, not its disability. The purpose of travel should be, indeed, to make one a *foreigner* and to force us to feel what that is like. And then to embrace that. This will not happen if we go only to Disney-controlled environments.

In the *Oxford English Dictionary* one of the words grouped in the definition of *foreign*, along with *alien* and *dissimilar*, is the word *irrelevant*. I like this. I can think of nothing more difficult for my ego than to think of myself as irrelevant. Yet that is exactly what I am when I am walking through the streets of some foreign town, in Dar es Salaam, say, or Mexico City. It matters not, then, to the hordes of other people on the street whether I am a traveler or a tourist. I am confronted at every turn simply with my irrelevance. It is not my life that is happening around me on all sides; that life would be happening with or without me. I am a foreigner. Yet it is exactly this that I must embrace. I like the idea of the humility we must submit to every time we are forced to do that, every time we are forced to understand ourselves to be irrelevant. If humility, to paraphrase Benjamin Franklin, is the doorstep to wisdom, how blessed we must then be. When we have no power, we have no choice but to turn to a kind of childlike trust. To be a foreigner, wrote Sallie Tisdale a long time ago in an article titled "Never Let the Locals See Your Map," "is to be both conspicuous and invisible; you can't blend in completely, and you won't be recognized either. To be a foreigner is to surrender somewhat. It requires a certain kind of trust, a willingness to play the fool for awhile."[3] Or as Flaubert put it, travel makes one modest, you see what a tiny place you occupy in the world.

The main purpose of travel should be not to perpetuate the life we know, but to get us out of ourselves. Travel, and possibly love affairs, dish up the main great adventures available to most of us in all our lives. These two, uniquely, provide our chances to turn ourselves over to something or someone largely beyond our control. When we embark on a love affair, we enter into an adventure of discovery, discovery of the other and discovery of ourselves. We enter a world of wonder, high in risk because of the likelihood we will get in over our heads, but full of promise too because of what we are likely to discover even when we are over our heads. Travel, as John Steinbeck postulated, is like that too. The trip takes us, and the most exciting thing is that it takes us to where we most certainly could not have predicted we would go and into the arms and the care and the hospitality of those we could barely have imagined to exist.

Reminding us of our vulnerabilities is one of the bittersweet beauties of the travel experience. We are reminded of how much we must depend on other people, of how much we don't know—how much we don't know yet, and how much we will never know. When I travel I am constantly at the mercy of strangers, and the great, edifying surprise of it all is how well placed my trust turns out to be most times when I have no choice but to place it.

A memory that sticks with me firmly is of getting off the train from Paris to Barcelona in the middle of the night. This was in the early 1980s, not long after the death of Franco, and Spain was still a dark and medieval kind of place. Outside the train station, it was a wet and chilly January. I was young, traveling with a friend, and since we had made no plans we felt we had little option but to take the offer of one of the hustlers on the station platform. In due course we found ourselves following through a labyrinth of darker and ever more ominous streets. Eventually we arrived somewhere—we had no idea where we were—and reached into our pockets to hand over eight or ten dollars so we could be taken into a glum-looking building and down a hallway past some rough-looking card players raucously jabbering in a strange language. Cigarette smoke and the heavy odors of cooking sat like some unwelcome uncle in the airless halls. I felt I was being led into a dream from which I might never emerge.

We were given a room: one that was about the size and shape and temperature of a refrigerator. The ceiling was high above our heads, and the walls crowded in so close to the saggy bed that we could barely move around it. A bare light bulb, hanging high above us, flickered a shadowy light. We settled in and agreed that my job for the rest of the night, as my friend attempted to sleep, would be to sit guard at the door. I have no idea, in retrospect, whether the fear that kept me awake and alert all night was merited. Probably it was not; in the morning everybody and everything, even the grim hallway, seemed quite benign. The strange language I couldn't place with the aid of my Berlitz Spanish phrase book was, of course, Catalan: my friend and I were very young and knew next to nothing. But the point is that rather than stumbling into bad experiences, in my travels I have, by contrast, been offered incredible hospitality, safety, and assurance by total strangers who wanted nothing from me in return other than a conversation, a tale about my country, or perhaps the opportunity to bestow their kindnesses. I have been taken on a duck hunting expedition by cowboys encountered in a restaurant in a small town in the middle of Guanajuato state in Mexico, and on safari by a camp owner in Zimbabwe. In both cases, my hosts were people I had known less than half an hour before the offer was made. Hospitality always, in my experience, is more available than trouble.

A number of philosophers have tried to explain the urges behind our travel as an inbred need in the human species to "wander." It seems that sitting still might not be what several million years of evolution have prepared us for. A famous book in the 1980s was Bruce Chatwin's account of wandering across the outback of Australia, *The Songlines*, in which—largely through his encounters with Indigenous people—he was forced to meditate about the nomadic as opposed to the settled life. Without deciding clearly whether the wanderer and the settler are opposed parts of each of us, or opposed parts of our society—some of us wanderers, some of us settlers—Chatwin declares himself unequivocally predisposed in favor of the wanderer. He blames most of the ills of history, including aggressions starting with Cain against Abel, on the settler impulse. The wanderer impulse, by contrast, is freeing:

> As a general rule of biology, migratory societies are less aggressive than sedentary ones. There is one obvious reason why this should be so: the migration itself, like the pilgrimage, is the hard journey, the "leveler" on which the fit survive and stragglers fall by the wayside. The journey thus pre-empts the need for hierarchies and shows of dominance. The "dictators" of the animal kingdom are those who live in an ambiance of plenty. The anarchists, as always, are the "gentlemen of the road."[4]

One of the charms of Chatwin's book are the pages of copious notes he assembles referencing thinkers with opinions on the need to keep moving. One is seventeenth-century philosopher, Richard Burton, who wrote *The Anatomy of Melancholy* (1621). Therein, Burton cited travel as a cure for the depressions brought on to the human populace by settlement: "There is nothing better than a change of air in this malady [melancholia], than to wander up and down, as those Tartari Zalmohenses that live in hordes, and take the opportunity of times, places, seasons." He linked the human urge for movement to the patterns already established in the cosmos and postulated that movement must thus be ordained: "The heavens themselves run continually round, the sun riseth and sets, the moon increaseth, stars and planets keep their constant motions, the air is still tossed by the winds, the waters ebb and flow, to their conservation no doubt, to teach us that we should ever be in motion."[5]

He goes to the Danish theologian Søren Kierkegaard for further advice on the mental health benefits of ambulation. Kierkegaard, in 1847, wrote to a friend, "Above all, do not lose your desire to walk: every day I walk myself into a state of well-being and walk away from every illness; I have walked myself into my best thoughts, and I know of no thought so burdensome that one cannot walk away from it . . . but by sitting still, and the more one sits still, the closer one comes to feeling ill . . . Thus if one just keeps on walking, everything will be all right."[6] Ralph Waldo Emerson wrote about judging the state of one's health—both physical and mental—by the number of walking shoes one had worn out.

There is an optimistic side to all these views of travel. I once knew a former priest who had started an organization that touted world peace as the end product of travel. As the numbers of travelers and destinations increased, so, he believed, should global peace. How could you go to war with somebody you'd just visited? Keeping on top of the news, we know that things have scarcely worked out that way—there is a long way to go. Yet we can always hope.

* * *

This is the point at which we switch from thinking about how, traditionally, travel changed us and turn to consider how it has come to pass that we, through our travels, now more and more affect and even change not ourselves but the world in which we are traveling.

On the southwest coast of Thailand, there is a spectacular inlet called Maya Bay. As I write this, Thai Forest Department biologists are painstakingly at work, as they have been for four years, planting coral in a project to slowly rehabilitate the bay. They also hope sea life will return. This comes more than twenty years after a Hollywood movie, *The Beach*, starring Leonardo DiCaprio, made the location so globally famous, tourists "loved it to death." Once the movie came out in 2000, there was an avalanche of tourism; the number of day-tripper tourists increased five- or even tenfold from a thousand per day to eight thousand and more. And, consequently, everything went wrong: the activity of the tourists and their boats destroyed over 90 percent of the bay's coral reefs. What had been beautiful and what had attracted everybody was, by that activity, wiped out. In 2018, the beach was, as their last resort, completely closed by Thai authorities, and the attempt at rehabilitation was begun. The costs of this work landed largely on the Thais; after years of trying in the courts, they had managed to squeeze only $273,000 out of the film's producer, Hollywood's 20th Century Studios (formerly 20th Century Fox), though the original lawsuit asked for ten times that amount. This was a nightmarish end to things since, ironically, the storyline of *The Beach* is about a group of hippies trying to connect with something authentic.

It had, likewise, now become time to examine travel through a critical eye. Though people have always traveled, about fifty years ago an anthropologist at the University of California at Berkeley moved to designate tourism as a field worthy of its own sociologic study. Whereas anthropology had up until then been concerned with picking apart exotic prehistoric societies, Dr. Dean MacCannell told his students that an equal field of study could be the citizens of "modern" societies who were roaming the world in increasing numbers just looking at things. Or seeking experience. If it was academically legitimate to observe and study so-called primitive societies, why not employ the same tools to have a look at a rather intriguing manifestation of our own societies—the enterprise of mass tourism? Immediately the heavy-laden language of sociology was brought to bear. Suddenly such issues as the "authentic" experience or sightseeing "ethics" became part of the vocabulary.

MacCannell's pronouncements were, in many ways, less than kind— or optimistic. One of his main observation-based critiques concerned what he took to be the superficiality of the mass tourists' interest and the smugness of their approach. He weighed in on the debate about tourist versus traveler with not much good to say about either. "In some instances tourists test themselves by traveling into war zones or to remote 'primitive' cultures, precisely to risk exposure to acute differences," he wrote. "These tourists are called 'travelers.' Tourists are called 'tourists' in the pejorative sense for their failure (really and truly) to experience even minor differences."[7]

"Moralizing commentary about tourist versus traveler," he went on,

> remains a stubborn feature of literature. The assumed superiority of self-described travelers is based on their claim to have accessed the inner workings of the places they have visited; that they actually crossed the line, they did not just press their noses up against it. This difference between tourist and traveler, if there is any, has been exaggerated. Tourists and travelers in the world of moral difference, are equal in that they are both extended a kind of honorary infantile status when it comes to local normative demands. They are equally recipients of a special kind of demeaning indulgence when it comes to the efforts to learn a few words of a local language, or some simple dance steps. Both tourist and traveler are sometimes overpraised for even small cultural accomplishments. And, to the relief of their hosts, both eventually go home.[8]

The impulse for us to judge or comment on other people and their travels goes back probably to what the Egyptians might have thought when the Children of Israel left for the Promised Land. Or, certainly, to English villagers weighing in on their neighbors' weekend getaways to the sea via Thomas Cook's rail excursions. What are they doing? Why are they doing it? Who do they think they are? Fascination and derision alike abounded. And, as travel and mass tourism has become more and more popular, so has the need to evaluate and assess. Trip envy, trip braggadocio, trip disparagement have all become popular by-products as the activity of tourism and mass tourism has grown, then exploded. Good-natured—and sometimes acerbic—debates erupted between the celebrants and the critics. American historian Daniel Boorstin, known for his bons mots, had a go at scorning tourists wanting what he said were elaborately contrived, superficial pseudo-experiences. Others countered that, au contraire, tourists were better than that and, concerned about "authenticity," roamed the world in order to put themselves into the presence of the "real thing." Some argued that tourists only want to get away from home and it doesn't matter what they visit as long as it provides a change of routine. A few, proponents of newly conceived moral tourism, observed that not all was mindless, that many tourists were more noble, wanting to be good, or at least not bad.

Wearied by the snarky nature of much popular comment, a British lecturer at Christ Church Canterbury University was applauded when he pushed back with a 2003 book, *The Moralisation of Tourism*.[9] Jim Butcher, in penning this diatribe book, had had enough of the moralizing of the moralizers. Give people a break, he said. Tourism had become "no longer an innocent pleasure, it had been interpreted and reinterpreted as an activity which is ultimately damaging to receiving cultures and the environment." Yet those who made judgments would allow no reprieve. "New forms of tourism, such as ecotourism, alternative tourism, community tourism and ethical tourism," he wrote, "have been presented as morally superior alternatives to the package holiday, yet ironically, even advocates of these new, ethical tourism brands are increasingly subject to criticisms, not dissimilar to those they themselves level against package holidays."

Reviewers found his views refreshing: "The only book on the market to provide a sustained critique of conventional mass tourism's own critics, Butcher offers a counterpoint to the moral rhetoric steadily turning travellers into guilty tourists."[10] His take on things struck a chord with mass travelers or budget travelers who, going all the way back to Thomas Cook, were fed up with the snobbishness of wealthy elites who, being able to afford to go anywhere or do anything, felt theirs were the only activities that meant anything.

Where are we left, we who like to travel? Should we be ashamed after being scolded by professors like Dr. MacCannell? When we examine ourselves, do we find that we travel for fun and entertainment? Do we travel in order to educate ourselves? Do we travel for psychological reasons, to feed flagging spirits? Does it matter if we seek "authenticity" as opposed to the controlled "performance," that we traverse the Atlantic in a berth on a merchant ship as opposed to a Carnival Cruise liner or book into a farm B&B as opposed to interacting with cartoon characters in Disneyland? Are our hopes for "authenticity" banal? Does it make any positive difference to anybody but ourselves should we undertake "do-good" tourism and lay a few adobe bricks to build a medical clinic in some jungle as opposed to shelling out for the adventure of zip-lining over top of that jungle?

Could we possibly want to replace actual physical travel by simply turning our avatars loose in the Metaverse?

And then there is the matter of "no surprises." On one level, the appeal of safety and predictability is understandable: One would not want to board an airplane if there was a fifty-fifty chance of it not being able to complete the flight. But it is a double-edged sword. Predictability does something spooky: It turns us from actors into consumers. Shop, choose, and devour. Swallow and be done with it. We have been seduced on many levels into being "consumers" and I, therefore, appreciate the complaint that we are consumers rather than citizens. We consume education; we consume healthcare. We ought to resist any notion of becoming mere consumers of travel. There are few things left wherein we can flirt with the unknown; travel ought to remain one of them.

CHAPTER I

My friend Neil was a traveler. He spoke several languages including Spanish and Portuguese, and boasted about having been to more than fifty countries. Neil was one of those adventurers who had no aversion to moving about in battered taxis, unheated buses, slow and crowded trains. But then, a couple of years ago, he came down with a degenerative disease. Incurable. Eventually on his deathbed he, one by one, invited his friends for a round of last meetings. The discussions were frank—but cheerful. Neil was not a religious man but was sanguine about discussing death. He had a palatable analogy at the ready. "Well," he explained to me concerning what he expected to happen next, "I've always liked traveling." This would be another trip.

CHAPTER 2

Cuba

The Idiosyncratic Island That Bet Big on Tourism

WHAT HITS YOU IS THE AIR, LIKE A PUNCH IN THE FACE: FRAGRANT, dripping with moisture. It is night outside the terminal and, looking away from its dim lights and those of the idling buses and taxis waiting to transport the mob of new arrivals to their various hotels, the ink-black sky has the soft quality of velvet. But it is the air that you embrace, realizing you'll need to swim through it. This is why we come, when up north it is the middle of January.

The islands of the Caribbean are littered with tourist resorts and cruise ship ports. But among them, the sternly Communist nation of Cuba remains intriguing. Significantly squeezed by the strict economic and political embargo placed on it more than sixty years ago by its powerful neighbor, the United States, Cuba grasped early on that tourism from those parts of the world not controlled by the US—places that included not only Russia and China but Canada and Italy—might be one of the few routes open for it to inject foreign exchange into its desperate economy. So, bowing to the invitation and having bought our tickets, there we were, moving from arrivals area to bus and then into the black night on our way to a week of sand and sun on the beaches of Cuba.

Cuba is a plucky little country that, during its history, has faced every problem under the sun. It is currently a poor country. This is not because its population of 11.4 million is not intelligent, good natured, and hardworking; Rather, the factors of Cuba's material poverty lie in a complicated maze of ideological and geopolitical issues. The 1959 revolution was built on the premise of rooting out corruption and inequality and, while Fidel Castro instituted universal healthcare, free education, and housing for all, he likewise dispossessed thousands of market-economy capitalists and property owners. This provoked an exodus of refugees and garnered the eternal enmity of the United States—ninety miles away across the Strait of Florida. After attempts to upend the revolution through everything from invasions (Bay of Pigs) to various sorts of subterfuge, the Americans resorted to the crippling embargo on trade and travel that remains in place sixty years on. Seeing its own opportunity, Russia promptly—though with strings attached—stepped in to partially fill the breach. But two decades of their support ended when the Soviet Union collapsed in 1990, at which point Cuba was left orphaned and floundering.

Most economies function by having something to trade and an unshackled populace who are free to be inventive and entrepreneurial. Cuba has a deficit of both. The island country is ranked 113th among the world's export economies, with its main products being sugar, tobacco, nickel, coffee, citrus, shellfish, and medical products. As for the entrepreneurship of its people, on the plus side you can argue that Cuba is well organized. The problem, however, according to critics, is that it is over-organized. In 2000, 76 percent of the Cuban labor force was employed by the state, a situation that has changed only marginally in the two decades since.[1]

The two phenomena—the socialist model with strict controls on everything from initiative to dissent on one side, and the unfriendly Americans on the other—feed one another with the consequence that in Cuba eleven million people live below what the developed world considers the poverty line. It should not be thus: Cuba's agricultural sector produces sugarcane, potatoes, rice, tobacco, coffee, beans, citrus fruit, and livestock. On another level, Cuba's technology is impressive. With its high levels of literacy and education, it is no surprise that the country excels in the industries of biotechnology and pharmaceuticals despite challenges of investment capital. (Cubans produced their own COVID vaccine almost as quickly as did AstraZeneca and Pfizer in the industrial

West, and promptly got 90 percent of their population injected). Seventy-four percent of Cubans use the Internet, which is a sign of the population's eagerness to reach out into the broader world. Yet material goods—frequently including food—are at a premium. Cuba is predominantly a cashless society; the average monthly wage in 2020 was $150 USD. A voucher system and things like state-controlled rents cover the essentials but, with the materials embargo, even should Cubans have cash, there is no abundance of goods to buy other than—as many will tell you—on the thriving "black market."

According to the World Atlas of 2021, Cuba's top export partners are Venezuela, which accounts for 33.5 percent; Canada taking 15.9 percent; China taking 9.5 percent; and the Netherlands taking 4.5 percent of total exports. According to the same World Atlas, Cuba is the eighty-ninth among global importing economies. The country's top imports are food, machinery and equipment, chemicals, and petroleum. Its top import partners are Venezuela, China, Spain, Brazil, and Algeria. For fifty years, Canada carefully engineered its way around the US embargo, managing for the most part to do business without unduly riling its powerful neighbor—and by far its biggest trading partner. Canadian exports to Cuba during 2021, according to the United Nations COMTRADE database on international trade, were $226.68 million USD. One of the biggest foreign investors in Cuba with interests in nickel and cobalt mining and gas, oil, and electricity (and with 500 MW of capacity, it is the largest independent producer in the country) is a Canadian mining company, Toronto-based Sherritt International.

A noteworthy factoid is that sugar from cane was once a Cuban trade staple, but now physicians have taken its place. Cuban-trained doctors go all over the world, most famously to Venezuela in exchange for oil. Another noteworthy factoid is that, as you would expect in such a situation, Cuba has a thriving black market built, in part, on the hard currency that makes its way into the country via tips to workers in the tourism sector.

THE TOURISM FIX

The first place I went to stay as a tourist to Cuba in the early 2000s was a resort called Jibacoa. Cameleon Villas Jibacoa is a rambling village-style resort of sixty or so semidetached bungalows, all of poured concrete under gracefully curving roofs with patio porches in front. They are connected to one another, to a large cafeteria restaurant, and to a swimming pool area by paved walkways under soaring royal palms. At the end of the property in a small hall with bench seats, we would gather every evening to listen to Cuban music, watch Cuban dancers, or listen to Cuban jokes. The adjacent sea, with a sizable reef, is attractive mainly to divers and snorkelers but treated with indifference by sand worshippers because of its being quite rocky. The shoreline is also in sight of a large petrochemical plant with a flame-shooting stack. What is most interesting about Jibacoa, though, is that, about an hour's drive east of Havana, it was built in the 1970s not at all for commercial tourism as we understand it today, but as an R&R spot for Russians serving their time as part of the Soviet-Cuba support pact. Wanting to get out of their offices or industrial plants or off their military installations for a few days of rest, this is where the commissars came.

Along the road running parallel to the coast from Jibacoa, amid small scrabbly farms, is another feature of revolutionary Cuba: holiday camps for ordinary citizens. It might have been—and continues in some respects to be—a cashless society, but one thing the revolution entitled Cuban people to, along with healthcare, education, and housing, was regular holidays. Hence these camps, a number of which I had a look at. They were set up with varying levels of utilitarianism through to some with high comfort. Most had barracks-style accommodations, though some featured small cottages, volleyball courts, baseball diamonds, and beach access. They were available for Cubans on holiday, just as the Russians next door had been on holiday and, latterly, the Canadians from Quebec and Newfoundland and Toronto were on holiday. At the time, this was a stringent sort of segregation; it wasn't until well on into the 2010s that Cubans too (although only those from the upper echelons of society who have foreign currencies in their accounts) were permitted to pay their money and muck in with foreign tourists at the country's all-inclusive resorts.

Cuba has always attracted tourists. The German polymath and founder of modern science, Baron Alexander von Humboldt, landed there in 1809 as part of his extensive transatlantic research voyage. Stepping ashore at the east end of the island, he expressed shock at the contradiction between the stupendously beautiful flora of Cuba and its plantation economy built on the enslavement of Blacks.

By the mid-twentieth century, one could argue that the Castro revolution was spurred in part by public distaste for the predominant tourism of that day—foreigners, mostly American, drawn to an island overrun with mafioso-operated casinos and all the sin and exploitation that went with them. This was the public face of Cuba from 1920 to 1958, a country where the US-based mob had sufficient sway to lobby successfully for things like the construction of a tunnel under Havana Bay to make it easier to drive between their operations in Havana and what they had going in the hundred-kilometer-away beach resort of Varadero. Along with chasing the Mafia out, there were signs early in the revolutionary period that things would be different. Fidel Castro announced that in revamping the economy, bringing in free housing, medical care, food subsidies, and education, he didn't want "an island of bartenders and chambermaids" (though in many ways that's what the modern all-inclusives have produced).

What this meant was a period of virtually no tourism, and Cuba looking like an Eastern Bloc country. In the 1970s and 1980s, 85 percent of Cuban trade was with the Soviet Union. In terms of imports, only 5 percent of what they got was from what was labeled "the free world." When the USSR collapsed in 1991, it was a terrible shock, and a stunned Cuba entered what it termed the "Special Period." Scrambling to come up with a whole new plan, Castro and his advisors put forward three areas of development they believed would be needed for the country to survive, which were outlined in lengthy speeches Fidel delivered, including one in 1992 to the National Association of Small Farmers.[2] The "three pillars," as they were called, consisted of a food program to provide for domestic food needs; a technology program, which included development of the pharmaceutical industry and high-tech medical equipment; and finally a tourist program.

The tourism initiative was a strategy by the Cubans rather than a response to any demand from abroad. As anybody does when coming up with a business plan, the strengths and weaknesses were looked at, the predicament the country found itself in was assessed, and the decision of what gambles to take was made. The government opened the doors with an advertising campaign targeting all Western countries except the United States and printed brochures featuring girls in string bikinis that exclaimed "Come and be seduced."

Tourism options took a variety of forms, from urban stays in Havana, Santiago de Cuba, or Trinidad and Cienfuegos to far-ranging camping and bicycle trips through the countryside and to a number of eco-sites. But the bedrock, turning out to handle 70 percent of all visitors, would be the all-inclusive resorts, defined places where visitors would be offered a controlled package of sun, sand, and relaxation. This is now an industrial-level business that peaks during the fall and winter when the northern climates are cold and gloomy.

The system devised forty years ago and honed in the interim would take coastal lands, the prime of them being along the twenty-one-ki-lometer-long peninsula of Varadero; build resorts that include upscale accommodations, swimming pools, restaurants, and entertainment stages; and have foreign visitors come to participate in packages of tightly con-trolled experiences. All the land on which these resorts were constructed was owned, and continues to be owned, by the Cuban government; the resorts, according to the plan, would be operated by and marketed by for-eign companies from Spain, Mexico, Brazil, and Canada. Carefully vetted Cubans would staff the resorts—and take home the tips issued in CUC (Cuban convertible currency) or US dollars.[3] A winter season would run from November to April, with intensive travel via charter flights to and from a host of regional airports. In the office of the Cuba Tourist Board on the third floor of a commercial building in midtown Toronto, Lessner Gómez Molina, Cuba's representative in Canada, proudly pointed out to me all these airports on an expansive wall map of his country and provided an update advising me of the resorts that are operating and a number more that are under construction. It is an optimistic business.

The Cubans understood the capitalist tools of promotion and advertising. A Canadian video producer, Sherali Thaver, told me how in the 1990s he was hired numerous times to travel across the island shooting footage of natural scenery and cultural sites that he distilled into twelve-minute videos to be shown in five languages—Spanish, Portuguese, English, French, and German—on the television sets in all the rooms of fifty-three hotels owned by the government-run Cubanacan chain. The chambermaids were instructed to set the TV channels to the language of the particular guest and leave them running.

Tourism as an economic savior, however, turned out to be an up-and-down ride. According to Cuban government statistical data and WorldData.info, tourism revenues in 1995 amounted to $1.1 billion USD, or about 3.6 percent of the gross national product (GNP). About 745,000 tourists that year spent roughly $1,477 per person. By 2019, before the outbreak of the COVID-19 pandemic, sales tallied $2.65 billion—more money but a lower fraction (2.6 percent) of the country's GNP. Spending per person had also dropped; each visitor in 2019 spent an average of $619 on their vacation. In 2019, Cuba had 4.28 million visitors, but the whack of the COVID-19 pandemic set them back on their heels. By 2023 objectives had moderated considerably to the point where they were going to be happy to see 2.5 million guests.

Coincidental with this, Cubans, in the years since 2008, witnessed efforts to loosen and diversify their tightly controlled society. In 2016 the first tentative steps were taken to introduce elements of a market economy. Initially, 127 areas of economic activity were identified as an experimental gesture. Along with hundreds of thousands of small farms, Cuba's non-state sector was then to be composed of small private businesses run by artisans and tradesmen. With that innovation, around 600,000 people, about 13 percent of the workforce, joined the private sector. You could be a barber, a tire repairer, a palm tree trimmer, or "a dandy," as the government referred to Cubans who dressed up to pose for tourist photos. In 2021, the list was vastly expanded to more than two thousand different fields. A significant part of this was in the tourism sector, where big moves to permit privatization were allowed for taxi drivers, small restaurants, and home-run B&Bs called *casa particulars*. Now, if you google "casa particular" or Airbnb Cuba, you will land on literally hundreds of possibilities of rooms you can rent for a night or a week from the domicile owners.

Politically during that period, substantive changes took place. In 2008, Fidel Castro, the founder of the revolution and the country's leader for forty-nine years, stepped down from Cuba's presidency, handing things over to his seventy-six-year-old brother, Raúl, who had been head of the army. In 2016, Fidel died. Raúl remained in charge until 2020, when a man who hadn't even been born at the time of the revolution, Miguel Díaz-Canal, became president.

THE ATTRACTION OF CUBA

One thing about Cuba, "the Jewel of the Antilles," is its being—like other Caribbean destinations—an island. Psychologically, an island is the perfect destination for those seeking escape, wanting to take a break from the hurly-burly of regular life, and enchanted with the idea of being, at least for a time, barricaded from "all that." Cuba, with its geopolitical isolation, promised the epitome of such isolation. This was so seductive that I confess to being exceedingly ambivalent when I found, on my sixth or seventh visit, that CNN was suddenly available on the television in our room. CNN was the last thing I wanted, yet, as our week went on, we got seduced and incorporated Wolf Blitzer into our daily routine. Something in retrospect I felt diminished by.

Another important aspect of Cuban travel is the shadow of ideology. Immediately after the revolution, while commercial tourism was essentially shut down as "bourgeois," large groups of visitors sympathetic to the revolution arrived to be of assistance to that revolution by undertaking such tasks as helping with the sugarcane harvest.[4] Ever since, according to Joseph Scarpaci and Armando Portela in their 2009 book *Cuban Landscapes*,[5] a substantial tourism driver has been visitors wanting to make their statement about the US treatment of the island, or deliberately choosing a holiday place where they were unlikely to encounter very many—if any—Americans. "Anti-Americanism and Cuba are a perfect match. Like David and Goliath," they wrote after asking: "What is the magnetism for the international tourist regardless of the merits or demerits of the Communist government?"[6] It is an interesting and charged question. There are people in my own acquaintance who choose to visit Cuba for ideological reasons, as there are others who, for the same reasons, refuse to go there.

Were you to categorize tourism in Cuba more specifically, you would come up with half a dozen permutations. One of them, of course, would be sun and sand. The main notion behind the model of the all-inclusive resorts is the offering of sand and sea to winter-weary northerners from Canada and Europe with a side option of visits to colonial sites and cities. Another permutation is cultural. Cuba's present-day museums and lively art, music, and dance are certainly attractive to the foreign visitor, and the tourism business goes out of its way to make it possible for visitors to get access to these. Almost every resort has evening shows where the guests are treated to the arts of traveling troupes of government-paid musicians, singers, and dancers. Then there is historical tourism. If you want to revisit colonial days, you can do so through the medium of the architecture of everything from the massive fortresses with their rusted-out cannons put up by the Spanish in the mouth of Havana Bay, to cathedrals and other colonial buildings. Many of these old buildings are under stress from years of low maintenance, but most of them are likewise invaluable classics.

Eco-tourism is increasingly on the menu for backpackers, cyclists, and nature groups taking in the flora, fauna, and sea reefs. After that, "nostalgic tourism" is a big hit with those who want to transport themselves back to a time when a 1957 Chevy with big fins, bench seats, an oversize steering wheel, and lots of chrome could get you around. That or a convertible Ford Fairlane. For those with this appetite, Ernest Hemingway's various drinking haunts, half-naked showgirls at the Tropicana, and a kind of sanitized version of the swaggering nastiness of Lucky Luciano, Bugsy Siegel, or Fulgencio Batista can still be indulged.

Others, of course, will want an antidote to this, and can do so by throwing in their lot with "revolutionary tourism," learning about, or paying homage to, the Castro revolution and Cuba's struggle to stay independent and make it all work. In centers like Havana, Santiago, and Santa Clara, one can see the icons of the revolutionary fight: the *Granma*, the yacht that brought the eighty-some revolutionaries across the sea from Mexico in its crowded hold; the airplane that bombed the American invaders at the Bay of Pigs; the rocket that shot down a US spy plane. In Havana, the monstrously supersized Revolutionary Square is surrounded by the current government ministries emblazoned with the iconic likenesses of various stars of the big enterprise. In Santa Clara, in the middle of the country, a pilgrim site is the massive mausoleum for the remains of Castro sidekick Che Guevara. In short order, there will likely be equivalents around the country for Fidel, who died in 2016. For those interested in the revolutionary lifestyle, it is possible to have a look at collective agriculture, the medical clinics in every village, and the creative system—in a country with a dearth of private cars—of organized hitchhiking where functionaries in blue uniforms compel private automobiles to take in a couple additional pedestrians if they have room in their backseats.

A final category to add to the list is likely "romance and sex tourism." We will discuss this a bit more thoroughly when looking at the "negatives" of the tourist encounter, but suffice to say, sex and romance is a motivator for numerous people who travel, as it is an enticing option or opportunity for their hosts. And doubtless, Cuba functions on this level as much as does anyplace else.

One thing that defines tourism in many locales but not least so in Cuba is its effort to reimagine history. We live through stories and so everything needs to be fashioned into one. When a story gets told often enough or gets repeated firmly enough, it becomes a mythology to live by. Cuban tourism has worked hard to present the pre-revolutionary period of its story in a way that is enticing—or at least nonthreatening—while then showcasing the revolution in as many ways possible. In this narrative, the "bourgeois" pre-revolution becomes a necessary and logical intellectual precursor to the revolution itself. But there are things to watch out for: Is there a confusing contradiction in the offering up of historic old Havana, the bars and haunts frequented by Ernest Hemingway, the

Tropicana nightclub showgirls, the Buena Vista club music, and then the revolutionary monuments all in one go?

ARE THERE PROBLEMS WITH TOURISM IN CUBA? OR SIMPLY CONUNDRUMS AND IRONIES?

The same week in early 2023 that my wife and I landed for a holiday on the beaches of Varadero, not far away a makeshift boat carrying nineteen young Cubans capsized in a heavy sea in the Straits of Florida. The Cuban coast guard pulled five bodies out of the water and reported that a dozen more were unaccounted for. The boat had, in the middle of the night, left Cardenas, the coastal town just south of Varadero famous for being the closest point of land to Florida. And then it had come to ruin. The tragedy of this accident at sea was that it was not an extraordinary event, but a feature in a regular traffic of attempts by young Cubans fed up with the limitations of their country and enticed by dreams of something different.

To be a tourist in the middle of such tensions puts one in a tenuous spot. Where do we visitors fit within the sociopolitical dramas of our hosts? Do we just have our fun blindly oblivious to the pain and struggle of those around us? Are we pawns in some game being played without our awareness? The contradictions are thorny. Which brings us to our series of important questions: Are there, concerning tourism in Cuba, actual *negatives* facing the country and its population, or just conundrums, contradictions, and ironies? Cuba's fleets of sixty-year-old automobiles, kept in immaculate condition, are both a tourist draw and a reminder of the country's embargo-imposed shortages. Can baseball, food shortages, medical and pharmaceutical innovations, and tourism all be parts of the same society? In Old Havana, is its rehabilitation being done for the benefit of the Cuban people (some of whom were reputedly moved out of the area because they got in the way) or to set it up to be in harmony with the aesthetic expectations of the tourists?

More questions: For people like the policymakers of Cuba, is tourism a compromise of revolutionary principles? Even though it remains the government that receives the bulk of the income from the tourism enterprise, it might seem natural for socialist societies to oppose tourism should they see it as a venture embodying market capitalism. Yet in numerous developing societies, any notion of this is ignored as tourism has become viewed as a cornerstone of their economies. Commissars of tourism use revolutionary heritage sites to draw admiring tourists and also to hard-sell their version of their story. Ethnographer Florence Babb, in her 2011 book *The Tourism Encounter*, tells of seeing a handmade billboard in Havana that read "to defend tourism is to defend the Revolution."[7] Will tourism and its growth prove a boon for the socialist economy, the revolution? Or will all those visitors ultimately destabilize that revolution and bring it down. Already, the link between tourism and the rebirth of small businesses throughout the country—the aforementioned taxi owners and drivers, private guides, casa particulars—has been established, as has the socioeconomic disruption that happens when workers transition out of certain professions (like teachers or medical workers) in order to become tourism employees. This they do in no small part because of the access to tips.

The macroeconomics of tourism is another field engendering elements that might lead to an understandable puzzlement. A Cuban all-inclusive is one of the cheaper options for travelers looking for such things as a less expensive holiday, probably 20 to 50 percent less expensive than its equivalent in Aruba or the Dominican Republic. If this price-cutting is done for reasons of competitiveness, then it is understandable. But is such price undercutting sustainable? Who, along the chain, takes the hit? The booking agents, the airlines, the hotel companies, the Cuban government must all be sure they get paid. Then food and liquor for the visitor's stay has to be purchased and supplied. Is the wiggle room then to be found in the wages of the tourist staff—the gardeners, waiters, chambermaids—all paid a reputed pittance in salary (by the standards of any other society) but doing the jobs in the expectation of tips?

When in Cuba in 2023, I booked a day trip to Havana. It was a bus with an onboard guide who took forty of us on the two-hour drive to and then an energetic walkabout around Old Havana. This was followed by lunch. What I got to thinking about was the economics of the trip. There were forty of us at $75 USD per head. So there was $3,000 USD, or $4,000 CAD, on that bus. Where did the money go? It was a nice, comfortable Yutong coach, part of a fleet of the Chinese-made vehicles run by the tourism enterprises throughout the country. The driver and Leon, our voluble guide, were probably paid no more than $20 USD each, if that, though the tips collected at the end of the trip possibly gave them $100 each. Lunch, in the government-owned restaurant, might have had a base cost of $5 per head, with the serving staff and musicians who entertained likewise hoping to collect tips. Some of us spent a few pesos in cash for trinkets or bottled water from vendors and small shops. But my larger point is that very little of the $3,000 or $4,000 that bus collected actually went to the goods and labor that we were afforded for the day. So: where did it then go? A little would have been collected by the Canadian agency that booked us in, no doubt. The bus would have required some fuel. The remainder, I suspect, a great deal, went into the pockets of the Cuban government. This is a capsule of the economic model of tourism in this socialist country. Is it sustainable in perpetuity?

Another issue with potentially negative implications has to be the disparity between host and visitor. It's difficult to be a Canadian or a European in a country where it is commonly made known that the monthly wage of the locals is about half of the daily wage in the country you have come from. One wonders to what degree Cuban workers understand that many of the guests at their resorts are, in fact, not rich people but pensioners or middle-class working people on a saved-for holiday: electricians, retired teachers, civil servants. Or maybe that is of no consequence; it is a challenging reach to come to terms with when the very fact of one's being able to travel in the first place implies resources well beyond those of any but the very top tier in the host country. This disparity is an issue in many "developing" countries—perhaps actually less so in Cuba due to its socialism wherein the government makes certain to get its take from the tourism industry and in turn provides more realizable goods (healthcare, education) to its citizens than is the case in many comparable countries. But in too many places, the matter of the local populations simply being cheap labor to serve foreigners in resorts they themselves could never afford to frequent is indeed a thorny matter. Later in this book we will look at the term *leakage*, which compounds the issue of where, in the end, the economic benefit lands.

Then there is the potential negative of the commodification of culture. Havana in both its colonial and post-colonial—but pre-revolutionary—iteration was a city filled with beautiful, classic architecture. The wear and tear of two generations of low maintenance, plus the assault of natural disasters like hurricanes thundering in over the Malecon seawall, have left street after street of these gems in sad shape. Some of the once-grand structures have been knocked down completely, though many, in recent years, have been taken on for restoration. The scaffolding and signs of rehabilitation are everywhere. The inconsistency comes from the information that instead of such restorations being done for the benefit of the citizens who live in Havana, many of these folks, in fact, got moved out of the center of the city because they were an eyesore or a nuisance. They were seen as standing in the way of the building of a phantasmagoric set piece to suit the preconceptions of tourist visitors. To what degree are we, as visitors, shown not what *is*, but what is determined we might believe should be? Like Disneyland though less blatant. And what price does local "real" culture pay for this sleight of hand?

And then there is sex. In assessing the conjunction of sex and tourism there are many factors, lots of them far from straightforward. I was once given a magazine assignment in 2000 to accompany a group of about a dozen American men on a trip to St. Petersburg, Russia. The purpose of the trip? Hunting for wives.[8] On the buyer side there was, at the time, a myriad of such enterprises operating via the internet, where ostensibly lonely fellows could peruse hundreds of photos and read short bios of comely ladies not only in Russia and Eastern Europe, but in Asia and Latin America. For a price, one could open correspondence and then, for substantially more, sign up for a tour.

On an evening in November 2000, one such tour group left JFK, New York, for St. Petersburg via Helsinki. During our ten days, we stayed in a St. Petersburg hotel along with a convention of figure skaters. Every third night we would pile onto a bus and the lot of us would travel to a champagne party in a fancy hall where the Russian women outnumbered the visiting men—who by now had been joined by some Germans and Englishmen—by about five to one. Dancing was done, acquaintances were made, dates for the following days were set up. This carried on until the ten days were up. Everything happened—one would be naive to think there was no sex. Did money change hands? Were dinners bought or gifts exchanged? There was enough substantial romance that by day nine, paperwork was available for a few who had got so far as to want visas for their very new girlfriends to travel to the United States.

Whatever it was—sex, romance—it was also tourism. Visits to the Hermitage Museum were undertaken, souvenirs were purchased, dinners and drinks were paid for in abundance. Was there anything wrong with it? Sex tourism is a definite negative when it involves exploitation of any kind or, presumably, when local community mores are transgressed. Prostitution of minors anywhere in the world is a definite no-no, as is recruitment for child porn or for other versions of sex trafficking. But what about activities between consenting adults? Is the freely entered exchange—even that built on possibly long-shot hopes—anybody else's business? There are other prices to pay. Another magazine story I worked on in 2002 involved a Canadian woman who had traveled to the Dominican Republic with some girlfriends, met a local fellow, fell for him, and paved the way for him to come to Canada. What came with him to Toronto, causing a scramble in the public health community, was a case of drug-resistant infectious tuberculosis.[9]

In Cuba after the 1959 revolution, Fidel Castro attempted to promote a wholesome society, and in doing so tried to end prostitution. Sex workers, or *jinteras* (literally, jockeys or riders; male and gay prostitutes are *jinteros*), who had operated in the good old days were re-educated to become seamstresses. However, when the country opened for tourism thirty years later, the official attitudes relaxed. Sex was employed liberally in the tourism ads, with pictures of women wearing simultaneously a bikini and a come hither look. But was that "bad"? If a portion of visitors to Cuba want to hook up, and from the Cuban end there are willing hooker-uppers, is that a dreadful thing?

Florence Babb, doing her ethnographic research for *The Tourism Encounter*, would occasionally position herself in Havana's cafes and bars to try to catch out sexual deal-making. When she interviewed both the locals and visitors, many were coy, though some were completely up front about their game. Babb sometimes comes across as a stern or censorious chaperone who will brook no hanky-panky. An anthropologist like Babb is properly watchful and wary of the power imbalance between well-off visitors and materially strapped locals, but approbation for the sexual motivation in travel or holidaying is tough to sustain. The world's oldest profession is difficult to hold down—along with its place in the increasingly prevalent black market that is the parallel economy of a country like Cuba. Also hard to hold down is the human urge to hope and dream. Not unlike the case for the Russian women in St. Petersburg with the visiting Americans, for Cuban women (and some men) "sex and love provide not only a diversion and relief from economic problems but also a bit of hope . . . to be swept away by romance to new places and new lives."[10]

REAL TOURISM

The prolific post-colonial and Orientalist scholar Edward Said was critical of tourism, viewing it as an imperialist practice built on rendering non-Western societies from colonial to post-colonial times exotic. This perspective acknowledges that the visitor is in the driver's seat. He/she has the money and can call the tune. He/she has the power to come and go and, often, fails to behave with humility as a guest always ought to. He/she is in a position to boss things around just as colonials have done since time immemorial. It is through his/her eyes that the locals get defined. The degree to which that is, in fact, the case is worth considering.

When you stay in an all-inclusive, almost all of the locals you meet are the resort staff: the waiters, the baristas, the bartenders, the cooks, the gardeners, the housekeeping staff, the front desk receptionists, the entertainment staff. The visitor, in this position, can hardly claim to get insight into "ordinary Cubans." Such forms of employment, mundane in many economies, are, in Cuba, much sought after and cherished. Such employment is, first of all, a steady paycheck with the added benefit of contact with a constant turnover of foreigners who have, among other things, the means to leave tips in hard currency. You can imagine what it means to be a tour guide who, in a country where the average monthly wage is $150, after a daylong shift of escorting a busload of foreign tourists on an outing from their resort to Havana, pockets $100 in tips.

To achieve such employment one needs everything from language skills to political connections. Being good at changing the sheets on a bed is only the beginning. Staff need to be seen as safe by whoever is in authority above them, going all the way to the top. They will not be exchanging any radical ideas with the visiting foreigners or asking or answering any politically loaded questions. Along with changing sheets they are, in fact, ambassadors of the Cuban ethic. All are thus, either by nature or by training, unfailingly cheerful and courteous. When you walk along a pathway the gardeners will stop whatever they are doing to say *hola*. Then back to pulling weeds or planting bulbs. There is likewise the pressure to self-improve and be ready. A waiter we got to know in one of the resorts on Varadero left every Tuesday to go by bus to Havana to study Chinese. The Chinese were on the map as potentially the next wave of Cuban tourism and Gustavo, who was about forty and already spoke Spanish, English, French, and a little Italian, was going to be ready when they arrived.

The relationship between the staff in the tourism sector and the hordes of visitors who are their raw material is a delicate and interesting one. In his study *Culture on Tour: Ethnographies of Travel*, Edward Bruner paints locals not as passive to tourism invaders from the outside, but as engaged in a co-production.[11] Which is an enticing premise as far as it goes. Yet what does it actually mean? You could argue about co-dependence, though in a sense there is a co-dependence in every mercantile transaction: somebody wants something and the other person can use what is offered in exchange. There is also, as I have witnessed in various Cuban visits, certainly to all-inclusives where people stay for a week or more, the development of relationships that are more than commercial between visitors and staff. People come to the same resort sometimes a couple of times a year and, when they arrive, tearful hugs are bestowed on favorite waiters or chambermaids as if what is taking place is a family reunion. Sometimes gifts are brought. But doesn't the co-production extend all the way through the system, even to the hustlers who attach themselves to, say, a walking tour of a group of Germans on the streets of Old Havana? The fellow on crutches, the young woman with a plaintive story about her hungry baby, the young man with a music CD or a charcoal sketch to sell, the middle-aged woman dressed in bright period costume and willing to pose for pictures, the teenager wanting to cadge a pair of sunglasses or a ball cap from a sympathetic visitor in exchange for a packet of chewing gum? These too are players in the co-production; they are in their locations, ready to go to work every day, hope springing eternal in their attitudes, rejection taken philosophically. Each understands that this is their economy, and that in the greater economy they too play a role.

The co-production at its highest levels certainly fell apart when the COVID-19 pandemic landed in the early months of 2020. My wife and I were holidaying in a resort on Varadero at the time and, spurred by the sudden alarm spread over CNN and by everybody else, scrambled to get to the airport and the safety of "home." On our crowded bus to the equally crowded and chaotic airport, the young tour guide could not stop weeping. The uncertainty of the moment and of the future overwhelmed him. When, he sobbed, might he ever get a job like this again?

We don't know what happened to that particular young man but, three years on, I am back chatting with Mabel. Mabel works late afternoons and early evenings in one of the niche bars at the resort my wife and I visit in 2023. Mabel can mix and cheerfully deliver any drink under the sun so long as it includes coffee. She started working in the resort industry in 2010. When the pandemic shut everything down, she "went home and sat in my house with my kids." She watched some people die, then got vaccinated with the home-grown COVID vaccine invented by Cuba at about the same time AstraZeneca and Pfizer showed up in Europe and America. In early 2022 she returned to work. "The government paid us a salary for six months," she says. "Then that was that." Mabel was out of work for almost two years. She was secure in her housing; her village had its medical clinic. She says people got sick but, if statistics are to be believed, Cubans died at much lower rates than in either Canada or the United States. Mainly, she had to forego the extra cash of her tips.

Our resort, one of the BeLive group owned by Spaniards, was closed from March 2020 until June of 2021. According to Tanya, who works behind the reception desk, tourists came back slowly and the workers likewise returned in dribs and drabs. The first Canadians, the main-stay group of this particular resort, came in November of 2021. But it remained a slow process: In the spring of 2023, it was about 60 percent occupied, while the resort my wife and I had stayed at in March of 2020 had yet to reopen. All this is to point out how dependent not just the Cuban economy but a segment of real, live Cuban workers are on tourists coming to their island, and what a blow it was when the flow was abruptly interrupted. It was doubtless the same for the street hustlers in Havana.

CUBA NEXT

The truly delicate moment affecting the balance of sustainability for Cuban tourism depends—as do many other things in that country—on the Americans. It is no secret that in a lot of eyes, Cuban tourism lives for the day when the Americans will come. Expectations lurch along in expectation of the island being fully open to the 340 million potential visitors only a few hundred kilometers from its shores. Twenty years ago, Americans could only get to Cuba by traveling via a third country, Canada or Mexico. Then during the Obama presidency in 2016, after a presidential visit, significant opening happened, including an exchange of ambassadors and the easing of travel restrictions and restrictions against transport of money to and fro. All this was rescinded in 2017 with the Trump presidency. In 2023, the United States restored full consular services in order to facilitate migration and travel. But the prohibitions of the Helms-Burton Act of 1996 preventing US-based companies from doing business or investing in Cuba remained in force, and so long as that is the case, there will be no great influx of money, investment, or attention from America, as those who hope for that dramatic change so long for.

Of course, there is the downside and worry. If and when they come, the Americans will swamp tourism in Cuba as they swamp everything else. Cuba will no longer be the same for those who come from other countries. Already in 2023 new hotels are under construction as one enters the Varadero peninsula—sprawling, grand affairs several times larger than many of the older places. As one Canadian who has been coming for many years declared, "I'll have to stop coming here when it turns into Las Vegas."

Las Vegas or Atlantic City is what the US mob made Cuba in the two decades prior to the revolution. Does anybody want that to be the case again? The irony is that Americans coming in large numbers as tourists would likely signal the end of the Cuban Revolution. It would mean that the long-standing embargo had been lifted. It would push the now-nascent market economy into much higher realms. It would remove the one strong card still held by large portions of the Cuban population: their willingness to suffer materially in order to remain distinct from America. The question that bears thinking about and addressing is this: what kind of balance that will prove sustainable can be struck between the needs of the present and the potential excesses of the future? One might look at the other islands in the Caribbean that presently accommodate American visitors and entertain US-based cruise ships. Are they—Jamaica, Barbados, Dominica—better off?

Desiring a sense of what Cubans who are not in the tourism business might think about the present configurations, I ask a young man I encounter in Havana whether he thinks there are "too many" tourists. I am sitting in one of the many small green parks with a statue in its center and iron benches along its walkways. The young man and his friend come by and stop. They are well-dressed and friendly. One is a student, the other says he works in technology. Around us women are carting home their shopping, someone is walking a tiny dog, an elderly man in a badly fitted suit is asleep on the grass, young boys are kicking a soccer ball. "No," he says, "there are not too many tourists. But this," he adds, pointing to the ground he is standing on, "is the real Cuba. Not Varadero."

CHAPTER 3

What Happens When Too Much Is Too Much?

Over-tourism in Venice, Rome, Barcelona, and Granada

THERE ARE ALARMISTS WHO SPECULATE THAT WITHIN A DECADE—BY as early as 2030 according to some—the fabled, iconic city of Venice could be rendered a ghost town. That is, so far as its permanent residents are concerned; no one will actually live there anymore. Two scenarios are put forward. One, that the island that is home to one of the most enchanting cities on the planet will sink under the weight of its architecture into the ever-rising sea. Ankle-deep water already occasionally washes over the pavements of St. Mark's Square, better known as the Piazza San Marco, before receding back into the lagoon. The other scenario is that the city will be fatally overwhelmed not from water, but by tourists. Venice, in this picture, will be relegated to some sort of Disneyland, its piazzas, churches, and markets turned into a kind of weird theme park of both history and fantasy. There will, in this one, be no room for the "real" Venetians, the "real" locals.

The blueprints for both outcomes are already in place. While the region (the Veneto) with a population of 4.3 million and the city of Venice, in totality 260,000 people, have been standing pat, the Centro Storico, or the large island that most of us think of when we think of Venice, saw that its population of permanent residents had fallen, in 2022, below 50,000, less than half of the 120,000 of forty years earlier.

The population in the 1950s was 170,000. Resident (and self-declared surrealist anarchist) Matteo Secchi told the German daily newspaper *Frankfurter Allgemeine*: "If it continues like this, we will become a ghost town like Pompeii." "Mr. Secchi and supporters from his organization, Venessia.com, warned of a continuing exodus, dubbed 'Venexodus,' as young people grow tired of soaring food prices, increasing tourist numbers, high rents and going without cars."[1]

Either of Venice's potential fates would be epic in the annals of civilization's catastrophes. Described sometimes as an ongoing compromise between land and sea, Venice is truly a water city. The main aspect of both the regional and urban geography is the shallow, crescent-shaped lagoon fed by three rivers that flow down toward the sea from the Dolomites. Over time, silt, carried by those rivers, built islands atop the heavy clay bottom of the seabed, and Venice, the city, is constructed on 118 of these islands. People get about through 150 channels and canals that course under 400 bridges via water transport, the most picturesque being the historic flat-bottom gondolas, though, latterly, any manner of fume-belching motorboat. The water aspect of the city has always been paramount. For its first centuries, prior to the fall of the republic in 1797, an annual ritual was the Doge, or local ruler, performing a marriage ceremony of the city and the sea on the day of the feast of the Ascension in May.

From its beginning, twelve hundred years ago, Venice not only thrived but became rich and famous. In AD 828 some cheeky merchants managed to steal what were generally believed to be the remains of the biblical apostle St. Mark from their place of rest in Alexandria, Egypt. These they transported home, where work began on construction of the Basilica San Marco in order to house the relics. Riches grew upon riches. By the end of the tenth century and the disruptions caused by the First Crusade, Venetians set upon a course of taking advantage of any European or Middle Eastern geopolitical turmoil. They managed to turn all four Crusades from the tenth to the thirteenth centuries to their benefit, particularly the fourth. When, during that last escapade Constantinople was plundered, Venetians ended up with most of the booty—substantial items of which got deposited in San Marco, such as statuary for the interior and reliefs worked into the basilica's outer facade.

San Marco was originally a chapel for the ruler of the city-state, the Doge, whose palace sits next door. Not hereditary, though elected for life, doges were symbolic rulers responsible to an aristocratic council that was representative of commerce as much as anything else. Venice was always about trade: merchants took all the daring risks necessary to amalgamate their prowess and riches. In 1271 it was from Venice that Marco Polo set off for his overland trip to China, returning twenty years later and forever altering the configurations of the world. Venetian power was further consolidated by the defeat of one of its main rivals, the city-state of Genoa, in 1380. Venice was cosmopolitan as well as rich and powerful; Jews and Armenians were given protection and religious freedom. Style, beauty, and a sense of specialness continued as the city's artisan class focused on the making of such luxury goods as Murano glass.

Such an interesting, beautiful, and accessible city could not remain a secret. Venice grew prestigious in the eyes of tourists through the nineteenth century, when it lived as a destination on the grand tour of the socially prominent along with artistic Europeans. These moneyed and leisured visitors painted and wrote about the sights, and turned the city into a must-see destination. Exotic Venice has been a mecca for writers, travelers, and foreign tourism ever since Lord Byron took his daily swim in the Canal Grande and swanned about the cafes of the Piazza San Marco. Hugh Honour in his city guidebook describes the visits of the global outsiders one after the other: "Ruskin, busy with his water colours."[2] (Nineteenth-century writer John Ruskin was such an avid traveler that he reputedly refused to consummate his marriage to Effie Gray in 1848 using the argument that should his young wife become pregnant, it would compromise their travel plans.) Honour goes on to list everybody else: Nietzsche contemplating the pigeons; Goethe enthusing over his first view of the sea; Henry James, Baron Corvo, Byron; Dante surveying the Arsenal. Petrarch, Aretino, Galileo. Montaigne, Horace Walpole, and Edward Gibbon all found things to complain about—the water in the canals was often stagnant and stinking. James Boswell took his pleasure with the courtesans and prostitutes, while D. H. Lawrence pronounced Venice "an abhorrent, green, slippery city."[3] The list of notables who passed time in the city over the centuries goes on and on and on. Giuseppe Verdi, A. E. Housman, Robert Browning, Richard Wagner (who died there in 1883), James Fenimore Cooper, Marcel Proust, Henry Wadsworth Longfellow, J. M. W. Turner, future czars of Russia and emperors of Austria, William Beckford, the British novelist who later became Lord Mayor of London. And, of course, the author of *Death in Venice*, Thomas Mann.

That was then. Now: at the apex of the Adriatic it was natural that Venice, latterly, should become one of the major destinations on the Mediterranean cruise ship itineraries when that industry came into its heyday thirty or forty years ago. Starting in the 1980s—when cruise ships became bigger, cruising fleets became more extensive, and the passenger lists became, shall we say, less elite—Venice was one of the first places to feel itself overwhelmed. The ships came in and docked, the day-trippers spilled out, and the city reeled.

A stop in Venice, according to friends of mine who took a Mediterranean and Adriatic cruise a decade ago, went like this: John and Hilda had boarded their Carnival Cruise ship in Rome. Once out of Rome, they sailed on to Naples, Dubrovnik, and then Venice. As they arrived in the late afternoon, everybody scrambled to the upper and lido decks to view the city coming slowly into focus. Theirs was a medium-size ship with fourteen decks and about two thousand passengers. Photos saved from the trip are gorgeous, showing Venice on a clear, blue day. But even their medium-size ship still towered ten stories over the city. The photos they've kept are almost surreal, in the sense that they might as easily have been taken through the window of a passing airplane. From the other side of things, one can imagine what the looming ship must have looked like from the streets down below as it was bumped by tugboats into its berth.

Once docked, the gangplank was lowered and two thousand Americans, Canadians, Brits, and Germans galloped off to join the already-thronged piazza. It was late season, October, but all the ports on their itinerary were still busy, Venice among them. The passengers were all back on the ship for dinner and in their berths for the night, but then all two thousand had the next day on their own in the city, wandering, shopping, and undertaking the must-do gondola ride. It was beautiful to be there, Hilda and John reminisce. And they and their fellow travelers certainly felt they got their money's worth, especially with such an easy transition from ship to shore that could be achieved by simply walking off their boat and into the very center of the city.

However, from the other side, Venetians were coming to the point of not entirely liking this. In 2021, after a couple decades of assorted stress and pressure, the city pushed back. In a dramatic act, the Italian government announced on July 12 of that year that all cruise ships, except small, boutique ships and river vessels, were to be banned from central Venice and the existing cruise port. Some alternative, not yet quite identified, would be set up, but for the time being the hordes would be diverted away to give Venetians and the overstressed sites of their city breathing room. Cruise ships, following this edict, were to enter through the narrow channel past the Lido, head round the island of Venice, and then travel up the waterway to either the industrial port at Marghera or to other locations on the mainland. The government claims its long-term goal is to create a new artificial harbor and cruise port on the Adriatic at some unstated (and probably imaginary) time in the future. It is questionable whether the initiative actually lowers the number of tourists since, after docking at the alternative ports, tourists, just as in the past, still make their way back to and into Venice for their day-visits. San Marco will still get its throngs; the intimate laneways will still have strangers pushing up and down through them. But on some level, the gesture made a difference: No longer will there be the unsettling specter of fourteen-story ships—or even more immense ones—parked at the docks and towering over the architecture of the city.

One of the ironies about Venice, of course, is that posed by writer Mary McCarthy in her classic *Venice Observed* way back in 1956 in which she described Venice as a "folding picture post-card of itself." "There is no use pretending," she wrote, "that the tourist Venice is not the real Venice. . . . The tourist Venice *is* Venice: the gondolas, the sunsets, the changing light, Harry's Bar, Torcello, Murano, Burano, the pigeons. . . . It has been part museum, part amusement park, living off the entrance fees of tourists ever since the early eighteenth century when its former sources of revenue ran dry." Is it, in fact, anything other than a tourist mecca? The locals might argue otherwise, but the counterpoint remains compelling.

The tourist urge to visit Venice remains close to insatiable. In 2019, the last reliable year of numbers before the COVID pandemic, more than five and a half million tourists arrived—a 5.1 percent increase over the year before. This was an average of 15,132 visitors per day, and about 30 percent of the resident population. Numbers on weekends and in summer months were considerably higher. Not all, of course, came via cruise ships. But they did come, with the preponderance being foreign visitors rather than other Italians. According to the Venice Yearbook of Tourism, foreigners of late represent 86.5 percent of tourist arrivals, as opposed to 13.5 percent being made up of Italians. Of the totals, 15 percent were Americans; 8 percent from the UK; 10 percent from Asia; and 13 percent from other European countries such as Germany, France, and Spain.

All these visitors push up against the thing that catches in one's throat about Venice: its fragility. Venice, more than many places on the planet, impresses as a truly delicate entity. There is, certainly, the fragile beauty of its artworks and the delicacy of its buildings. But, more so, there is the fragility of its very physical existence. Venice, as alluded earlier, barely holds its own in its battle with the surrounding waters. The city, for one thing, is slowly sinking under its own weight (almost 3 inches over the past thousand years but 9.4 inches [24 cm] in just the last century). But you have to couple this sinking of the wonderful buildings of the city into their foundations with the fact that the Adriatic is simultaneously rising. In slow motion the world is watching Venice sink and the sea surrounding it rise.

As long as there has been a Venice, water has been not only its blessing but its challenge, the channels through the lagoon regularly silting up and needing to be dredged. At the best of times, the Adriatic tides will wash them out, but any that became landlocked quickly turned stagnant and the citizens would need to deal with malarial-level swamps. The clever Venetians built engineering works to continually divert the rivers through wide channels to enter the sea at various points from out of the lagoon. These worked with the incoming tides to keep everything fluid—a natural sanitation that carried refuse away and rendered Venice one of the cleanest cities in Europe.

Not long ago, however, when the rising sea became something Venetians could no longer ignore, they set about to install their biggest engineering project ever—massive floodgates for the mouth of the harbor. This project, designed to protect the city, would, in fact, serve to fix the lagoon in place. Begun at the turn of the millennium, the Modulo Sperimentale Elettromeccanico (MOSE) consists of seventy-eight barrier gates that, though not yet completely finished, faced their first test in October 2020 when their closing managed to at least partially protect the city from a 4.6-foot tide surge.[4] The year previous, a state of emergency was declared when Venice experienced its worst floods in fifty years, when six feet of water briefly inundated even St. Mark's.

It is in the context of such myriad pressures on the city's existential health that the Venetian city managers made the decision about cruise ships. The longer or even middle-term effects are still to be measured. Will the moves tend to discourage tourists in the long run? Will that satisfy any disgruntled locals? Does the decision ultimately signal a trade-off between tourism revenue and heritage conservation? Are other jurisdictions going to be influenced by Venice to try similar control measures where they live?

Venice is far from the only city in Europe, or on the planet, forced to rethink its relationship with the masses of tourists clamoring to visit. For almost a decade, the Greek island of Santorini has been trying to figure out how to discourage cruise ships. Due to the strains that over-tourism puts on the small island's resources, officials in 2019 set a goal to cut the number of daily tourists by 20 percent, from ten thousand to eight thousand. One way to do this, they believed, would be—as in Venice—to keep the cruise ships further away from the center of things. The upshot is that ships cruising the Greek islands now have to dock inside the caldera, the half-moon-shaped bay created by an ancient volcanic depression. From there, they tender the passengers some distance to Santorini's Old Port, after which these visitors require a further exertion to take a hike or a cable ride to get to the town of Fira. Officials are keeping their fingers crossed. "A lot of people here depend on cruisers, but something has to give," Santorini mayor Nikos Zorzos told the *Washington Post*. "The electricity grid and water supply are at their limit. Garbage has doubled in five years. If we don't control the crowds, it will backfire and ruin us."[5]

Viewed for such a long time as an uncomplicated "good," tourism is now undergoing cost-benefit assessments by numerous jurisdictions taking a look at both economic and environmental pluses and minuses. The results are not always positive. Both citizens and governments are questioning with more intensity and sophistication the degrees to which it is "worth it." When a cruise ship docks in the harbor and disgorges five or six thousand visitors onto a city's main square—there to mill about for four or five hours, have a cup of coffee, buy a few trinkets, and, maybe, have lunch before getting back on the ship for their main dinner meal, their evening entertainment, and their beds—what economic good does this do for the average Venetian? Or Montenegrin, Liverpudlian, or Bahamian? Increasingly, locals who have moved from patient amusement, to caustic jokes, to outright churlishness have come to see a whole raft of tourist activities as a nuisance. Sometimes a great nuisance. When locals want to go about their business, they have to negotiate their way through streets and squares crowded with foreigners taking pictures; when they want a cup of coffee, the cafes are full and the prices inflated. When they want to pray in their local church, chances are it will be crowded with more foreigners taking pictures. Then, instead of dropping any real money, the foreigners disappear back onto the boat which, in all of this, seems to be the only real financial benefactor. Oh, but the town council will get docking fees and that will pay for services and keep taxes down, will they not? Such arguments are proving a bit abstract to the increasingly irritated locals.

ROME

Anybody who has been to Rome either as a student, a backpacker, or a more formal tourist has probably spent time sprawled on the Spanish Steps, chatting with your neighbors, enjoying a gelato, or just imbibing the atmosphere. No more. In 2019, city officials elected to ban tourists from loitering on the Spanish Steps. Arguing that the move was necessary for both public safety and to preserve the landmark, Rome would no longer permit mobs of people to laze in the warm sunshine, have a coffee or a soda, canoodle with their lover, or chat with friends on the grand 138-step staircase that sweeps down from the church of Santissima Trinità to the Piazza di Spagna. A financial fine of up to 400 euros was designed to "guarantee decorum, security, and legality" as well as avoid actions that are "not compatible with the historical and artistic decorum" of the city.[6] An early step was to make eating and drinking verboten, but now it would be the mere act of sitting.

Presumably Romans as well as outsider tourists stand to suffer from the prohibition—since how would the distinction be made? But the underlying message was clear: It is the foreign tourists, in their millions, that were causing the overcrowding on this freebie of the Roman experience. Making and then enforcing a rule would be a step in curtailing their rampant freedoms and keeping them under control. Later, the prohibition was modified, allowing a fifteen-minute stay, but the principle remained clear: Keep moving.

Are such rules mere churlishness on the part of local governments? Or are they necessitated by what locals truly consider to be situations that are increasingly out of control? The operative word used to label the phenomenon of local frustration is *over-tourism*. This is a term that has not been used much up to the present, but now is ever more common. Use of the term *over* produces an unmistakable definition: at a certain point balances tip; what had at one time been the welcome embrace of interested visitors becomes a burden, an annoyance. The cost-benefit analyses show too much cost and too little benefit. It is a delicate balance that emerges from a typical pattern of officials and populations first noting it, then not knowing what to do about it, then becoming increasingly frustrated by impotence, then finally exploding.

When Romans grouse that their city is going downhill, they say it is "becoming like Venice." This truly dramatic slur is their way of saying they are scraping the bottom of the barrel. As is the case for the Venetians, tourists are the bugaboos, and there are many fresh annoyances having to do, certainly, with numbers but especially the proliferation of something that accommodates the growing numbers, Airbnbs. In a 2023 article in the Toronto newspaper *Globe and Mail*, journalist Eric Reguly, who lives in Rome with his family, relayed the concerns of many Romans that the city's historic center was on its way to being emptied of locals in favor of mass tourism.[7] There was hard evidence for this: The "real" population of Rome's historic center, according to Reguly's research, had fallen in three decades by 20,000, from 190,000 to 170,000. Neighborhoods are being transformed in order to cater to travelers. And again, the Airbnbs proved an identifiable culprit.

I, as a visiting tourist, might be delighted to land an Airbnb apartment just up the street from St. Peter's; the Romans, on the other hand, are not so happy to see me pricing that space out of the reach of them or their friends and families who are looking not for some place to holiday, but some place to live. What is good for tourists is seen more and more as no good at all for the locals. In fact, just the opposite. Universally, the locals see their lives becoming more difficult and more expensive. They blame the increasing numbers of tourists for noise, crowding, and a general atmosphere of disrespect. The proliferation of unregulated—or certainly not regulated up to a par that is considered satisfactory—short-term rentals are crowding out apartments that would have been available to locals; old neighborhood vegetable markets are being replaced by restaurants and bars "of the non-Italian variety." The old city is being turned into an "architectural theme park." And, the coup de resistance, Starbucks franchises have been set up and are selling frappuccinos, a drink that appalls Romans.

The last time we were in Rome, my wife and I stayed in a family-run hotel. Stephanie, who had lived in Rome when she was a teenager, said that the Romans always despised tourists, as if the attitude is imbued at birth. Still, as noted subsequently by Reguly, the tourists were a more obvious part of the scene than she remembered from back then. The shops she had been used to frequenting in the Monti district were fewer and harder to find. There certainly were Starbucks competing with the traditional coffee bars.

BARCELONA

Another European city to vent its frustrations is Barcelona. The capital of the Spanish region of Catalonia, Barcelona landed on the tourism map after it hosted the Olympic Games in 1992. At the end of the dreary decades of Generalissimo Franco, Spain was more than ready to become modern, and nothing spelled modern better than this rundown industrial port city giving itself a spruce-up. Barcelona soon became known not as the center of republicanism during the dreadful civil war, but as a city with some spectacular pieces of architecture—like the perpetually unfinished Antoni Gaudí church, la Sagrada Familia—and the festive street of cafes, Las Ramblas. Barcelona was simultaneously fast becoming one of the fashion design centers of Europe.

In light of this dramatic transformation, the city was eager to show itself off to incoming visitors from all over the world. In short order, however, the golden goose's egg started to make people nervous. Barcelona has a population of 1.6 million but was soon getting 32 million annual tourist visitors. All this led to a panic among its residents about the impacts of tourism. Rather than net beneficial, all the visitors might well be affecting residents' lives in negative ways.

As far back as 2016, citizens of Barcelona started to react through such moves as gathering on the city dock to greet the (then) world's largest cruise ship with signs that read "Enough of massive tourism. Benefits for just a few, costs for everyone" and "Let's defend our neighborhoods, our air and our health." The city's mayor since 2015, Ada Colau Ballano, Barcelona's first female mayor and an unabashed advocate of "the people," took the side of the complainants. The high demand for tourist accommodation, coupled with the opportunity to rent out rooms and apartments via sites such as Airbnb, meant—as in Rome and Venice—that rents were soaring and residents were struggling to afford the increases. Over-tourism, Colau was quoted as stating, "is affecting not only residents' quality of life but their very ability to live in the area."[8]

According to travel writer Vicki Brown, it took some time for Barcelona's officials to come round to such a critical point of view. Despite the city's "clear shortage of capacity," the local government and tourist board had spent years clamoring to have ever more visitors. One of the outcomes was a kind of chaotic business plan where, in the case of Barcelona, 125,000 legally registered hotel beds and apartments found themselves up against and competing with 50,000 illegal beds. The bureaucracy, in its zeal, had lost control. This wide-eyed scramble by the municipal tourist boards is not a novel phenomenon. As I write, I note on the news of how Quebec City in Canada is mounting a campaign to be the featured destination on an HBO television series called *The White Lotus*. Apparently the region's tourism arm, tirelessly chasing possibilities, is all in for this venture, as tourist boards seem to universally be.

In Barcelona, as in Venice, the cruise ship trade came under scrutiny. Of the 32 million annual visitors (2017), 2.7 million were of the cruise ship variety, docking and making their mad dash up and down the Ramblas. This is facilitated by Barcelona possessing the Mediterranean's biggest port, but the numbers of visitors using that port had exploded exponentially from 115,000 in 1990 to almost 3 million seventeen years later. According to Brown's reporting, in 2019 Barcelona (along with Palma) also managed to earn the unwanted title of most polluted port in Europe.

Almost a decade ago, Dr. Harold Goodwin, founding director of the International Centre for Responsible Tourism, authored a paper titled "Managing Tourism in Barcelona" (2016). In it he made an observation about the city that summed up the dilemma of both the local officials and the local citizenry:

> Of the 32 million annual visitors, around half are day trippers; this will include most of the cruise passengers. Spending just a few hours in the city, visitors have a limited radius and will tick off the same few places: La Rambla, the Sagrada Familia, Parc Guell, La Boqueria market. [It all feels like] a tidal wave when after breakfasting on board the ship there are as many as 35,000 people arriving at the Mirador de Colon, to walk up La Rambla. Little money is spent outside of the main tourist cafes and souvenir stalls, and the congestion is unpleasant for both residents and other tourists. In the case of the cruise passengers, many will book tours through the cruise line itself which leaves even less money in the city.[9]

Barcelona was in the same pickle as Rome and Venice: too many visitors causing too much disruption with a negative cost-benefit balance.

GRANADA

Lastly, another iconic city with a tourism attraction every bit as seductive and profound as Venice is Granada in the south of Spain. The Moorish period, lasting about eight hundred years from the eighth century to the fifteenth, was created in the region now known as Andalusia by the Al-Andalus civilization brought in by North African and Syrian Muslims. In 1238, the emir Ibn al-Ahmar began construction of the stupendous Alhambra palace and fortress atop the region's Sabika hill. An aggregation of the Alcazaba citadel, Generalife residences, and a medina, or townsite, the constructions assembled by a series of rulers of the Nasrid dynasty remain cool, delicate spaces, witness to the elegance of this largest center of western Moorish culture.

The Moors held sway through a number of historical epochs until the Christian "Reconquista" in 1492, after which Ferdinand and Isabella moved in and used the Royal Court they had just taken over as the place from which to give Christopher Columbus his marching orders. Three hundred years later, as part of Napoleon making a mess of too much of Europe, segments of the Alhambra were laid waste by his armies. But sufficient was salvaged that, another two hundred years on, the site came under the umbrella of UNESCO world heritage and became a major tourist destination.

As a tourist destination, Granada faced the same conflicting challenges as numerous other places. How to show off the gems of its heritage without seeing them destroyed or at least damaged in the process? Also, how to make the economics of hosting multitudes of visitors sustainable for its own population? A recent study of these questions was undertaken by Canadian academic Elaine McIlwraith of Western University in London, Ontario. McIlwraith documented how substantial numbers of Granadians found at least some bits of income and employment catering to tourists. Among the more matter-of-fact aspects she considered were the levels of dependence of locals on the gig jobs of freelance tour guiding and Airbnb hosting. She concluded that these forms of income were important. She also determined that, not surprisingly, this exerted pressures toward expanding rather than controlling tourism. "Homeowners and tour guides in Granada, Spain often rely on tourism labour generated by economic models that promote continuous growth and result in over-tourism," she wrote. This had its built-in ironies and contradictions, and she spotted them. When anybody proposed limits to tourism, there was an automatic pushback. "Sustainable models proposed align with their desire for conviviality and governance yet call for consideration of their complex situation."[10]

The issue was complicated on many levels. Everybody, from freelance tour guides to homeowners who rented out rooms or offered Airbnbs, was happy to benefit from the economy of tourism. In the cases of some of the younger workers, being in the business, even on a part-time basis, allowed them to remain living in Granada rather than having to migrate elsewhere in search of livelihoods. But, on the other side of the coin, these same young people were simultaneously among those who worried about the effects of too much of it. In the abstract of her paper, McIlwraith writes that though the tour guides were happy to lead walking tours for visitors, they were alert to the fact that "this exponentially increased number of tours hinders everyday movement of residents." Homeowners who converted their homes into tourist apartments, which they themselves managed, were simultaneously aware that the "practices have disintegrated the community feel in the neighborhoods." The nub of people's conflictions came through the fact that "the same groups, younger tour guides and homeowners, that depend the most on capitalist models of continuous growth to get through, or recover from, periods of precarious employment, are often those that desire sustainable tourism and the implementation of strategies of de-growth." They, of course, she concedes, "may not be in a position to advocate for them."[11]

AIRBNB
Central to the complaints of disgruntled citizens about over-tourism in cities like Venice, Rome, Barcelona, and Granada, along with many others across the world, is the exponential growth of the phenomenon of short-term rentals of spaces other than hotels. Hospitality outside

of hotels and in people's homes has long existed in many societies through relatively small-time initiatives such as beds-and-breakfasts or family-run pensions. Even letting traveling relatives stay with you during their holidays could be seen as an informal part of the system. But it is only in the last fifteen years that a phenomenal mix of technology and enterprise has turned what used to be essentially mom-and-pop ventures into a juggernaut built on global online booking and reservations that mass all the small-time initiatives under corporate reservation umbrellas.

The biggest, most corporate, and best organized is Airbnb. In 2008, AirBedandBreakfast.com consisted of three young guys sleeping on inflatable mattresses on the floor of an apartment in San Francisco. Fantastically, this would grow into a $31 billion publicly traded company with 4 million hosts across every country of the globe accommodating 1.4 billion annual guests.[12] It was also to become the flashpoint in the mix of grievances held by local populations against their hordes of visitors.

I, along with many others, have been a consumer or user of Airbnb or its equivalents. My experience developed over a number of years. Early on, attending a conference in Croatia, I arranged, through an intermediary, for use of an apartment. This was not an Airbnb arrangement, though in form it was very much the same thing. My understanding was that I was getting a bedroom and bath that, for 20 euros a night, seemed a great deal. The location was not in the middle but in a scenic suburb of Dubrovnik, up a steep hill with a view over neighborhoods and as far as the sea. My hostess, an older woman who kept this apartment for her son who was away in Zagreb, lived next door. She met me with the keys and, on the morning after the first night, popped by to see how I was doing. Having sized me up and determined that I was somehow okay, she said, "Larry, please use the kitchen also." I thanked her and, on my way back that evening, purchased some snacks and staples. The next day, our relationship grew: "Larry, please use the living room." I now felt free to relax with the TV. By day four, it was: "Larry, please use the roof" and, suddenly, I had access to a terrace with a spectacular view out to the Adriatic. I was still paying 20 euros a night and felt like I had a home in Dubrovnik.

A couple of years later, in Lyon, France, my wife and I rented for a week from a "Superhost." Josephine met us in the lobby of her building and gave us our keys. The apartment—a third floor with balcony—was beautifully situated. An épicerie and a patisserie were just down the street, as were a number of cafes and restaurants. A metro station was two blocks away. We could make an easy morning walk to the beautiful Parc de la Tête d'Or, home of Lyon's botanical gardens. Were I to live in Lyon, this would be the spot I'd want to be. Josephine's apartment was spotless, with a superb coffee-making machine. It was, again, a semiprivate place, owned and used periodically by her mother, who lived primarily in a small village some distance away. There was a mild though not overpowering sense that this was somebody's home and we were visitors. There were some innocuous books on the shelves; a number of family photographs were present, though they had all been turned to face the wall so as not to intrude on the guests. We could still sort of assume it to be "our place" for the few days we spent there.

Then, more recently, on a holiday to Victoria, the capital of British Columbia, we went for the full Airbnb experience. We connected online with hosts Ken and Jennifer. Ken and Jennifer, for all I know, might have been fictions of some Singapore-based numbered company. We communicated only through a series of possibly AI-generated emails that instructed us what to do when we arrived, how to access keys, and so on. The apartment was fine, on the sixth floor of a newish building on a very convenient street. This one was not lived in by any host's son or mother. Furnished in standard IKEA, it was as comfortable and as inoffensively neutral as a hotel room. We discovered in short order that we were not alone as guests in much of the six-story building. Numerous among the singles, couples, and young families coming through the lobby and using the elevators were, like us, pulling their wheeled luggage and consulting their iPhones. The street outside was convenient to some of the main scenic sights of the city, and hosted a number of bars and restaurants. This made it all fine for us as tourists. It was as good a deal as the nearby hotels; we appreciated the freedom to keep a few items in the refrigerator, and it was convenient, uncomplicated, and well located. It never occurred to me to think about my role in the macro situation—or even that I might have a role.

All the waiters and baristas and dishwashers who worked the bars and restaurants on the streets below us in Lyon or in Victoria, where did they live? How was my presence affecting them? Was I taking up an apartment where one or two of them might otherwise have lived? On the other hand, without me or my ilk, would any of them have had a job? The developing line, however, is that for those who aren't benefiting from tourism by, say, being employed in it, my presence likely both priced and spaced them out of this housing market. And that makes them unhappy. This is the crux of more big pushback, the world over, by locals against over-tourism, and takes us back to Venice where, coincidentally, the Airbnb pushback story too has roots.

In early 2023, the *Times* reported how the mayor of Venice, in another (different from the pushback on cruise ships) attempt to avoid being overwhelmed by tourism, had confirmed plans to curtail Airbnb rentals: "Luigi Brugnaro said he was determined to discourage home-owners from letting their properties to visitors for more than 120 days a year. 'We want to stop the city and its lagoon becoming places for tourists only. We won't allow that,' Brugnaro told the Italian paper *Quotidiano Nazionale*. Brugnaro plans to charge tourists an entry fee and hopes that by limiting rentals to four months of the year, he can stem the tide of apartments being turned into hotel rooms."[13] Unlike cruise ships where the issue was too many tourists not leaving enough money or benefit behind, the accommodation issue pits private small businesspeople—like Josephine in Lyon or my landlady in Dubrovnik or "Ken and Jennifer" in Victoria who see a chance to make a few extra dollars—against the interests of their fellow citizens who see only rising rents and housing shortages. Add to this the hotel industry that sees a competition they consider to be under-regulated and unfair.

In the Venice story, Airbnb, resenting being singled out as a culprit, pushed back. An Airbnb spokesman in the *Times* story stated that "the majority of Venetians renting to tourists continued to live in their homes, and that 40 per cent claimed this rental income was what enabled them to stay." The spokesman added: "We welcome regulation and have put forward proposals for clear and simple national rules that support everyday families and clamp down on speculators that drive housing and over-tourism concerns."[14]

Such controls do exist in a number of places. Paris has a limit on rentals of 120 days per year; London caps things at 90 days and Amsterdam 30 days. (Amsterdam actually added to this with an active "stay away" campaign which, on the city's website, told would-be day-trip tourists not to bother showing up at all, especially if they were simply going to be drunk and noisy.) In 2022, Italy's national government passed a law allowing Venice to take action but, as of the *New York Times* story being published, the city was still formulating rules to make them legally sound. Other Italian cities reportedly keen to restrict Airbnb included Rome, where the population of the historic center has fallen by about twenty thousand in the past thirty years, and Florence, where about one in five homes is now a tourist rental.

For its part, Venice has continued to push things further. In late 2023, the city upped the ante another notch by approving a controversial 5 euro tax to be levied on day-trippers who come into the city center at the year's busiest times—Easter, for example, or summer weekends. Those who book an overnight stay will be spared, but the cheapest customers, the day visitors, will be dinged, the intention being either to enhance the city coffers or to discourage mobs of those who expend less than they supposedly ought to.

This sense of places being "overly touristed" is a transformational issue and one pleading to be sorted out. If at some point officials or communities come up with creative ways to address not only the housing and short-term rentals but everything else that bothers locals in tourism-heavy areas, there might be a good outcome. If not, who knows? There might be national laws; there might be bans. But what then to do with the tourists who are bound to keep coming?

CHAPTER 4

Cruise Ships

Blessing or Curse?

COUNTLESS PEOPLE LOVE TO GO ON A CRUISE. AND THAT IS WHAT THEY are doing, cruising—don't ever mistakenly call it "traveling." The only traveling might be to get to the port where the cruise loads. After that, cruising is "riding"; when you come back to land, it will more than likely be the same place from which you departed—in other words, you haven't gone anywhere.

Cruising—unless your ship should sink—is a safe, predictable, controllable experience. Instead of being on land dipping your toe into the water, as would tourists at a beach resort for example, cruisers are on the water occasionally dipping their toes onto the land. When you dock somewhere, it will only be for a few hours, enough time to mob the town, buy a few trinkets, annoy the locals, and then get back on board where the real business of cruising will continue: eating, sleeping, being entertained, maybe doing a little gambling in the onboard casino. And riding. From your balcony or out through your portholes, you will see water. And sky. When you spot land, it will be a coastline going by as if you are watching it on a movie screen.

Yet in the tourism industry, cruise ships occupy an enormous space. The website Cruise Mummy advises that in 2022, sixty cruise lines operated 322 ships worldwide. The industry is a moneymaker. The $27 billion in revenue in 2019 dropped dramatically during the pandemic, but was expected to more than recover to $30 billion by 2024. Number one in revenue is Carnival Cruise Lines, which earned $12.1 billion in 2019. In that year, there were fifteen million cruise passengers, a number that dropped to three million in 2020 and rose to five million in 2021 but again was expected to recover—or more than recover—by 2024.

The cruise industry is a huge employer. It employs people in offices and booking agencies; it employs workers managing and staffing the ports, workers maintaining and operating the ships, workers building the ships, and housekeeping, hospitality, food service, and entertainment staffs. The biggest cruise company, again Carnival Corporation, headquartered in Miami and incorporating a host of smaller companies and brands like P&O out of Australia, Princess Cruises, and Holland America Line, claims to have a hundred thousand employees. Carnival's ninety-plus ships touch in at over seven hundred ports worldwide. Its website boasts:

> a talented, passionate and diverse workforce of people from nearly 150 countries, and its brands host nearly 13 million annual guests historically—accounting for nearly half of the overall global cruise market. Combining over 225,000 daily cruise guests and 100,000 shipboard employees, more than 325,000 people are sailing aboard the Carnival Corporation fleet every single day, totalling about 85 million passenger cruise days a year historically.[1]

It might seem tedious to return here to the COVID pandemic of 2020. However, there is possibly no business on earth for which that international health emergency put things into such stark relief as the global cruise ship industry. The scope of its operations, the flaws in its structures, the changes that might have to be made for things to move forward were all highlighted and highlighted dramatically. As the pandemic struck in the spring of 2020, a prominent news story playing everywhere was of cruise ships and their passengers marooned at sea. Suddenly those holidays on the ocean, much sought after by their aficionados, became the worst place one wanted to be. With their closed quarters and limited onboard medical services, the ships started to look like floating incubators of disease and, when nobody knew quite what to do with them, they quickly became roaming pariahs. When COVID broke out on board, suddenly everybody had it. When the ships wanted to dock and let their passengers go home, no port would accept them.

In February 2020, Royal Caribbean's *Anthem of the Seas*, a 1,140-foot vessel with the capacity to carry 4,905 passengers and 1,500 crew, was quarantined in Bayonne, New Jersey, until guests who had recently been to mainland China were checked out for their flu-like symptoms. A sense of urgency was heightened as ambulances lined the docks at Cape Liberty Cruise Port. But the poster child for all the bad news was the *Diamond Princess*. This British-registered ship sailing out of Yokohama, Japan, and then back was the first to have a major COVID outbreak on board. Having left port on January 20, it quickly returned, hoping to unload. But instead it was denied docking privilege and was quarantined by Japan's Ministry of Health, Labor and Welfare. The quarantine lasted for about a month, from the 4th of February. During that time, things just got worse and worse. Of 3,711 passengers and crew, around 700 became infected, and 9 people died.

Around the world, governments and other ports almost instantly fell into line, preventing cruise ships that were out on the high seas from docking and ordering the rest of the industry to shut its doors. Cruising worldwide was totally suspended in mid-March. Still, ships already at sea were truly at sea. None were allowed to dock. By June, over forty cruise ships had confirmed positive cases of coronavirus. Marooned on board were hundreds of passengers and crew, all wondering when and how they would ever see land and home again.

The last ship to finally limp home was the *Artania*, which, after weeks of roaming, was finally allowed into the German port of Bremerhaven on June 8, where its last eight passengers disembarked. Back in March, the *Artania*, which was registered in the Bahamas, was anchored off the coast of western Australia, where health authorities confirmed seven coronavirus cases on board. Considered to be a foreign vessel, commonwealth forces were called in to refuel and resupply it before it was sent on its way. By March 27, forty-six people on board were reported displaying COVID symptoms. Forty-one passengers and crew who had tested positive were treated in private hospitals in the city of Perth. This left

450 crew and about a dozen passengers still on board as the Australian government ordered the vessel to move on. The captain resisted the order, arguing that the ship needed to stay in close proximity to healthcare facilities.

By April 7, two of the *Artania* passengers had died, and by April 23 they were joined by another passenger and a Filipino crew member. Finally leaving Australian waters, the ship, between April 18 and June 8, wandered from western Australia to Indonesia and then to the Philippines, dropping off 56 and then 236 crew members in those two countries. It then sailed toward Bremerhaven via Singapore, arriving after six harrowing months at sea with its last eight weary passengers.

A summary of cruise line statistical information published in December 2020, "COVID-19 and Cruise Ships," with help from the tracking website Cruise Mapper, detailed both how bad and how comprehensive the disaster had been. From mid-February until the end of March of that year, 3,260 passengers had been infected and 70 had died on 44 different cruise ships owned by 18 cruise lines. It was a dreadful tally.[2]

For its part, the adventures of the *Artania* underscored a host of problems no one had bothered to anticipate or foresee. In times of emergency, foreign-registered ships had few rights in other country's ports. Crews who hailed from all over the world had few protections in such emergencies. Countries had differing and changing rules about the welfare of crews, and since cruise lines often refused to pay, repatriation of crews became an enormous problem.

Yet, cruising was not an industry that could be kept down. Not only were memories of the travails short, huge numbers of potential passengers refused to give them a second thought when as soon as possible in the pandemic's waning, advertising campaigns set out to convince the public that they would now be safer and cruising was still fun. As early as the latter half of 2020, things were back up and running out of parts of Europe and in the South Pacific.

A typical promotion was posted at the time by Carnival:

Carnival Corporation's nine cruise line brands offer a broad range of vacation options for millions of guests with a wide variety of leisure-time activities that accommodate people from multiple backgrounds, cultures and languages. Based on Carnival Corporation's relentless commitment to consistently exceed guest expectations, the company's cruise offerings provide extraordinary vacations and exceptional value for travelers from all walks of life and for every type of vacation or occasion, especially when compared to similar land-based vacation options.[3]

Global cruising follows ordained cycles. Ships of the Atlantic, Caribbean, and Mediterranean fleet essentially ply the Caribbean in winter, departing out of Miami or perhaps New York. Then, in spring, these vessels head via Madeira, the Canary Islands, Lisbon, maybe Morocco to the Mediterranean, where they will work for the months of May and June and September and October. In the hot months of July and August, they take a break from the Mediterranean and head north into the Baltic and North Seas. Pacific fleets follow similar migratory patterns; ships that ply the coast of Alaska in summer can likely be found off Cabo or South America in winter, and possibly in the Far East in the shoulder seasons. Like birds, and some sea life, it is a migratory pattern both for the ships and for their hundreds of crew and personnel.

* * *

The beginning of systematic commercial pleasure cruising goes back almost two hundred years. In June of 1833, flying the flag of the Kingdom of the Two Sicilies, the *Francesco 1*, which had been built in 1831, left Naples loaded with nobility and royal princes from across Europe. This was an early example of cruising designed not to get anywhere, but simply for the pleasure of the ride and sightseeing. An advertising campaign had promoted stops along the Italian coast in Taormina, Catania, and Syracuse, then Malta, Corfu, Patras, Delphi, Zante, Athens, Smyrna, and Constantinople. The passengers apparently were not disappointed. It was reported that they ate lavishly, danced, played cards, and had deck parties. The *Francesco* was for snobs of the grand tour type—only the rich and the well connected on board.

Another decade passed before the P&O Line (Peninsular and Oriental Steam Navigation Company), founded originally as a mail service, started running voyages for a more general population. It would take these British passengers from Southampton to the Mediterranean and then on to stops in Constantinople and Alexandria. The P&O ship *Valetta*, built in 1889, was the first vessel to have electric lights. Cruising for the masses started to build from there. In short order, Thomas Cook got involved, taking his customers as far afield as Egypt, and by the 1880s had a fleet of forty vessels providing tours up the Nile.

A BUSINESS BUILT ON SUPERLATIVES

A few years ago, I arrived in Luka Kotor, Montenegro, by bus from Croatia, hoping to see the mountains. The first mountain I saw was a mountainous ship dominating the pier. It was ten stories high, had just turned in off the Adriatic, and completely overwhelmed the quaint little streets of the old town spread out like a rug at its feet. Going back to about the 1980s, when the activity of cruising ballooned, so did the size of the ships. Ever since, it has been one superlative after another.

The world's largest cruise ship in 2022 was Royal Caribbean's *Wonder of the Seas*, launched that year. It has the capacity to carry 6,988 passengers and 2,300 crew. It boasts not ten but eighteen decks, weighs in at 236,857 tons, and is registered in the Bahamas. The world's most expensive ship, built in 2022, was MSC's *World Europa*, costing $1.2 billion with twenty-two decks and a capacity for 5,400 passengers. In 2023, it would be eclipsed by Royal Caribbean's *Icon of the Seas*, which will cost $2 billion and carry 7,600 passengers. *Icon of the Seas* will be the world's biggest cruise ship, at 250,600 tons, and will be 6 percent bigger than its sister *Wonder of the Seas*. For the moment, ship building has stabilized: according to Cruise Mummy, eighteen cruise ships were retired in 2022 and seventeen new ones were added to the global fleet. But things probably won't stay calm for long.[4]

Up until the late 1970s or 1980, seagoing—and river-going—cruises were rather modest enterprises. The boats offered "deck chairs, shuffleboard, drinks with teensy umbrellas and little else for a few hundred passengers." After 1980, the ships quickly grew in size and heft to become as big and populated as midsize towns. They then competed with one another to offer every amenity and entertainment imaginable: The food to be served got more and more and better and better, the entertainment more and more lavish, and the onboard gyms and casinos and theaters better and better equipped.

When the shipbuilding boom commenced in the 1980s, the first cruisers to start to resemble ocean liners in both size and durable strength were the Sovereign Class. Built in Saint-Nazaire, France, at the Chantiers de l'Atlantique shipyards, these were considered the first mega-ships and could cross the Atlantic with no problem whatsoever. *Sovereign of the Seas*, launched in 1988, eclipsed an older ocean liner, the SS *Norway*, that until then held the record as the world's largest passenger ship. The *Norway* promptly responded by getting refitted with two more decks and, in 1990, reclaimed the size title.

As the new forms and styles caught on, the comforts and luxuries on board all new constructions exploded. Instead of ocean-view cabins, you now got entire decks of cabins with balconies. Instead of gangways, you had multistory atriums and glassed-in elevators. Recent iterations of cruise ships with their everywhere balconies are being described as floating condominiums.

These floating behemoths, however, carried their own built-in problems. They became targets for criticism and denunciation not only for their excesses, but also for their failures. According to their critics, cruise ships are responsible for terrible iniquities of many sorts. "Cruise ships kill whales, leak gray water, and are largely exempt from U.S. taxation. When they violate the law they pay the equivalent of a parking ticket," fumed Kim Heacox, an Alaska-based activist who had watched plenty of ships plying the coast of his home state, as he unloaded in the *Guardian* newspaper.[5] Such blanket-style accusations became common fodder. Writer Nichola Daunton, on the website euronews.travel, opined: "While a cruise around the Med [Mediterranean] or the Antarctic might seem like the height of all-inclusive luxury, you may not feel it's worth the negative impacts after you read this. New research reveals that passengers on a seven-day voyage around the Antarctic can produce as much CO_2 as the average European does in an entire year." The study she was referring to assessed that a big cruise liner produced a carbon footprint equivalent to that of 12,000 cars and that a night on board required twelve times as much energy as a hotel stay on shore.[6]

One outfit that is well organized in its criticism of the cruise industry is the Amsterdam-headquartered environmentalist network, Friends of the Earth. With bases in seventy-three countries after its founding in 1969 by disaffected Sierra Club members, Friends of the Earth pulls no punches on its website statement:[7]

> Cruise ships are a catastrophe for the environment—and that's not an overstatement. They dump toxic waste into our waters, fill the planet with carbon dioxide, and kill marine wildlife. Cruise ships' environmental impact is never ending, and they continue to get bigger. They once were small ships, around 30,000 tons. Now, corporations are building billion-dollar cruise ships to hold more than 9,000 people. They're doing everything they can to pack these floating cities full of tourists while polluting everything in their path.
>
> Unfortunately, everything that cruise ships come in contact with are likely to be harmed along their journey. The air, water, fragile habitats, coastal communities, and wildlife are all affected. But most governments have refused to take actions to actually regulate the cruise industry and buried their heads in the sand to ignore the ongoing damage to the environment and communities.

Friends of the Earth goes methodically through indictments of the big cruise lines one by one, starting with Carnival Corporation:

> Carnival Corporation is the biggest cruise company in the world with 10 cruise lines and is the most notorious cruise company for their environmental impacts. They have a long history of violating environmental regulations and have paid millions of dollars in fines—instead of cleaning up their act.

They've been charged with the following:

- Dumping food mixed with plastic waste in Bahamian waters

- Falsifying records of environmental compliance plans
- Illegally releasing over 500,000 gallons of sewage and 11,000 gallons of food waste globally
- Illegally discharging oily waste off the coast of England
- Illegally dumping thousands of gallons of wastewater into Glacier Bay National Park in Alaska

Carnival was hit with a $40 million fine for illegal waste disposal and was put on federal criminal probation in 2017. During their five-year probationary period an independent court-ordered monitor was required to examine its ships. During the first year, an inspector found over 800 violations. A year later, Carnival had to pay another $20 million for environmental violations. And in January 2022, it paid another $1 million for failing to implement court ordered monitoring programs to attempt to ensure it doesn't illegally pollute in the future.

Royal Caribbean comes in next as a wreaker of negatives:

Royal Caribbean's pollution is second only to Carnival Corporation when it comes to the criminal fines they've had to pay. They were forced to pay $18 million in fines for 21 federal felonies in 1999 due to dumping hazardous chemicals and waste oil in coastal waters. They then lied to the U.S. Coast Guard and Justice Department by falsifying oil logs to cover up their crimes.

Royal Caribbean has 25 ships in its fleet and 22 of them utilize scrubbers to get around community and climate-harming emissions. Scrubbers are used to "clean" smokestacks from the dirty fuel they use and reduce air pollution. But the scrubbers convert air pollution into water pollution because the toxic by-products from the scrubber systems are discharged into the seas.

At the end of 2021, Royal Caribbean announced "Destination Net Zero" to get ahead of the talks on green travel. But its targets toward sustainability are decades away and focus on carbon offsets instead of true sustainable measures. It will likely be too little too late.

THE SINS OF MSC CRUISES WERE THAT:

MSC Cruises is not an environmentally friendly cruise option either. In fact, they claim a range of environmental initiatives, but only do so half-heartedly.

Currently MSC Cruises has nineteen ships in their fleet, but:

- Only twelve have installed advanced sewage treatment systems
- Only eight have shore-side plug-in capability
- Only six travel to ports with shore-side power
- And fifteen utilize scrubbers that convert air pollution into water pollution

It's clear that MSC Cruises has a long way to go to minimize their environmental footprint and truly be a champion for the environment.

Disney Lines get off a bit lighter in comparison to the others, though concerns are raised about plans waiting on their to-do list:

Of the top cruise lines, Disney is open and transparent about their environmental impacts. Disney ships utilize fuel with a 0.1% sulphur content which is key to reducing climate-harming emissions. But Disney isn't perfect. They like to provide their vacationers with a once-in-a-lifetime travel experience, including going to untouched areas of the planet and to create a "private island" experience. And however idyllic that sounds, there's an environmental price to pay.

Disney is working on a massive cruise ship port in the Bahamas at Lighthouse Point. This port is opposed by community groups and will cause harm to coral reefs and migratory patterns of marine wildlife. The Bahamian Island of Eleuthera is known for its clear blue waters, sandy beaches, lemon sharks, and sea turtles. It also includes four endemic plant species and over 200 bird species. But Disney's planned Lighthouse Point port would bring a million visitors to this untouched region, adding noise, water, and air pollution to this habitat.

These indictments are both sweeping and detailed. They are critiques of environmental carelessness along with some brazen misdemeanors. They are very bad PR, coming at a time when sensitivities and vigilance among the public about environmental caretaking have reached increasingly urgent levels and are likely to stay there. The zeitgeist has less and less tolerance for improprieties and offers less and less wiggle room to get around them.

Complicating matters for things such as accountability is something that certainly came into focus during the COVID pandemic: Who exactly is it that can exercise governance and oversight for the vast industry? Ships, going back to pirate days, are notorious for skirting the laws of more-strict nations by flying flags of convenience from jurisdictions that are more easily manipulated. You don't want to adhere to the regulations of the UK or the United States? Register yourself in Panama or Liberia. You can spot the country of registry for any ship printed on the stern, just below its name. A ship belonging to a company based in Miami might be registered not in the United States but in the Bahamas. A flat fee is all that is required from certain countries to register a ship, no restrictions or onerous charges. And that gives it license. This, mainly done for tax reasons, can liberate a vessel from any number of other rules that might be imposed by more-stringent governments. The registration issue has frequently helped ships up to now to avoid things like onerous environmental regulations. But the tide is possibly turning—the appetite for cutting slack to big corporations is fast waning, leaving shipping lines needing to take a great deal more care with their behavior. Not just official watchdogs but, one would hope, passengers and potential passengers will insist on improved performances.

The critiques of cruising are not solely environmental. Another tier of complaint comes from their ports of call and, increasingly, from the citizens who live in those ports. Everywhere, from large, scenic cities like Venice or Barcelona (see chapter 3) to smaller islands in the Caribbean or the ports of Hawaii, if you want to get the take on cruise ships and their passengers from the standpoint of economics or culture, you only need to chat with the locals. Cruise lines all negotiate docking fees with the ports at which they call. A port fee is, as it suggests, a fee paid by a ship to dock in any particular location. The levies are assessed on the size of the ship (tonnage) and the number of passengers. They cover such costs as the pilot brought aboard to guide the ship to its berth, a rental of the berth itself for however long the ship will remain there, and a kind of "head tax" to offset costs of the passengers' use of the local infrastructure. These costs, of course, are buried in the fare each passenger pays for their cruise and are reputed to run from $100 to $500 per person.

This is income for the governors of the port and for the polis where the ship has docked. But—as explained in our chapter on Venice—if nothing has been negotiated with the citizens of those port towns, all the inconveniences and perceived nuisance can, in time, lead to rebellion. If they are asked to put up with inconvenience or any other sort of annoyance, citizens need to *see* a corresponding benefit. Should they be told that local hotels are doing well, it might not be enough to offset the fact that the prices they have to pay to get an ice cream cone are now inflated or that they can no longer get a seat in a local restaurant. This is a factor not just of cruises but all sorts of tourism, including in my own inland city where, when I encounter a mob coming down the street and crowding me over, I am impelled to ask: "What's in this for me?" The effect, though, is more dramatic when it is a cruise ship with six thousand passengers all spilling off at once. A friend tells about being in Venice (he was also a tourist, though not from a ship) and comparing the noise of the afternoon with the quiet of the evening—after the ships had departed. Some places are better than others at persuading their citizenry that any tourism benefits all local citizens. But as more time passes and as tourist crowds grow and as tourism becomes a bigger part than ever in certain economies, the pressures to do a better job on this front will only grow greater.

One doesn't have to look hard to find a barrage of additional criticism. The Business Insider website lists seven places being ruined by cruise ships and over-tourism.[8] Mallorca, Barcelona, the Great Barrier Reef, Santorini, Venice (six hundred cruise ships a year), Dubrovnik, the Galapagos make the cut. The account reports how eleven thousand residents of Mallorca signed an anti–cruise ship petition citing pollution, overcrowding, and waste buildup as their concerns.

MITIGATION?

It's not that the cruise lines are oblivious to the challenges they face. A number of improvements are either included in new vessels or in the offing for the very near future, many of them having to do with fuel and efficiency of engines. According to Cruise Mummy, more than 50 percent of all new cruise ships launched in 2022 and 2023 were powered by light natural gas (LNG) as opposed to diesel. This will produce zero sulphur. Is solar or hydrogen far off as a propellant? Ships, you might remember from the history books, were once powered by wind. Enhanced efforts at waste recycling is also on the menu. Royal Caribbean's *Symphony of the Seas*, for example, now boasts that it is a "zero landfill" ship; it uses recycling and water filtration to deal with its own waste entirely.[9]

On the other front, an innovation that feels recent (though it took its first tentative step in the 1990s) removes cruise ships almost entirely from the pressures they have been placing on populated ports. In order to avoid populated ports altogether and provide uncomplicated shore time for their passengers, major cruise lines have been buying up private islands. Mainly in the Caribbean, and mainly in a kind of arc around the Bahamas, these tiny atolls tend to have spectacular beaches while being otherwise basically unpopulated. Seven that are now the preserves of major cruise lines circle the Bahamas: Castaway Cay is owned by Disney, Half Moon Cay by the Holland America Line, and Perfect Day at Coco Cay and Labadee by Royal Caribbean International. Norwegian Cruise Lines owns Great Stirrup Cay and Harvest Caye; Princess Cruises owns Princess Cay. The drill is as follows: the ship moors, the passengers are ferried ashore and basically spend some hours enjoying the beaches before being returned to their ship. Some of these islands are very small; Labadee is only 260 acres, or 105 hectares. One hundred forty people live on the thousand-acre Castaway Cay and all work for Disney. But the islets all seem big enough to fulfill their stated purpose: sparing populated port towns some of the aggravations of too many tourists. The private islands solve that problem, but the program also deprives the ports and some of their businesses of income. So there's the double edge of the sword. The phenomenon, likewise, further isolates the cruise experience from being anything other than a cruise experience.

Finally, like some kind of postmodern irony, we get a weird phenomenon that has recently entered the picture called "cruising nowhere." If a cruise is a ride going essentially nowhere, this newish venture epitomizes the notion. Unapologetically—or without any sense of irony—the guests will be taken out onto the water, but will go nowhere. A cruising nowhere venture might last a week, or only a couple of days. Many have underlying purposes: They might be like spending three days at a spa, but a floating one. Or three days in a darkened casino, but a floating one. The enterprise has its fans. Various review articles waxed enthusiastic, arguing that, like staying in a nice hotel on land, you might have an enjoyable time without ever leaving your room. You didn't need to be anywhere else or going anywhere else.

The phenomenon has its own academic interpreters. Bradley Rink is a human geographer and associate professor in the Department of Geography, Environmental Studies and Tourism at the University of the Western Cape in South Africa. In an article in *Tourism Geographies*, he took a look at what he termed "a new generation of domestic South African cruise tourists," that is, on cruises to nowhere, "and the ways in which they perform liminality through a mix of sun, sea, sex and especially alcohol."[10] His review suggests a little bit of concern about the potential debauchery of the mix of too much "sun, sea, sex and especially alcohol." But it is an acknowledgment of a new sort of cruise holiday now being carved out.

English Cathedrals

Churches, Museums, or Amusement Parks?

A HALF-DOZEN YEARS AGO, ON A DRIVING TRIP THROUGH ENGLAND AND Wales, my wife and I visited a number of cathedrals. The first on the list was Hereford. Herefordshire is a county in the West Midlands bordered by Shropshire to the north and the Welsh Marches to the south and west. It is one of the most rural and sparsely populated counties in England, known for fruit, cider production, and agriculture. More cows than people, basically. The town of Hereford, with the River Wye running through, is a market town with an imposing statue of a big cow (doubtless a Hereford) on its main square.

Yet even more imposing than the statue of the big bovine is the grand Gothic cathedral. Begun in the eleventh century and situated on a hill above the river with the town of some sixty thousand surrounding it, this seat of the Hereford diocese of the Church of England is, in many senses, your garden-variety Gothic cathedral: solid, majestic, and forty times as big as anything else in the vicinity. But its being a bit out of the way adds to its iconic charm in that inside are stored some of the most valuable relics of the Middle Ages. These include the famous Mappa Mundi, a three-meter-square sheet of calfskin showing a fourteenth-century conception of the world, with Jerusalem at its center. The map's 500 drawings depict the history of the world up until then along with some natural marvels: 420 cities and towns; 15 biblical events; 33 plants, animals, birds;

32 images of peoples of the world; 8 pictures from classical mythology, all painstakingly etched into place.

Wanting to see all this, on a Saturday afternoon, Stephanie and I approached the reception kiosk and laid down £12. This gave us entry to the section of the cathedral where the Mappa is displayed as well as access to the Chained Library Exhibition that holds 1,500 rare books and manuscripts. It was an awesome journey into a thousand years ago. Two hours of browsing got followed by a visit to the Cathedral Café and a taking of our coffee and cakes to a table outdoors in the shady Chapter House garden.

The next day, Sunday, we returned. Leaving our lodgings, which were some distance out of town, and parking our rental car on Quay Street, we reentered the cathedral with no fee at all. This time it was because we were coming not as museum visitors, but as worshippers. Being Anglicans back home, we knew roughly what we were doing. Taking seats in the nave, we listened respectfully and meditatively to the organ prelude, then prepared to rise for the procession of choir and clergy and the beginnings of the service. We stood when everybody else stood, knelt when everybody else knelt, sang when everybody else sang, mumbled the various responses when everybody else mumbled the responses, and, when the collection basket came round, dropped in a tenner. All the while, though, we were stunningly aware of the majesty of our surroundings: the vaulted ceiling some sixty feet above our heads, the dancing light through the stained glass of the soaring windows, the looming presence of the high altar.

Our fellow parishioners in Hereford Cathedral—the big space had been more than half-filled—were warmly open and friendly, certainly by the standards of the English. At the close of the service, we were welcomed by several and urged to join in the after-church coffee that would be served in the gardens overlooking the river. There, among the profuse flowers of early July, we got chatting with a number of local folk, including the retired headmaster of the cathedral school. He informed me he was writing the history of the boys' school. This was interesting, because back in Canada I had just completed a project where I had been commissioned to write the history of a boys' choir school on the event of its fiftieth anniversary. "Oh," said my new friend. "That sounds an easier task than mine." The story of his school went back more than six hundred years. We stood then for a moment in silence. What more was there to say? In that information was everything: the reason we were in Hereford, the reason we were in England. History, and almost unfathomable spans of time.

In the course of one weekend, Stephanie and I had had both a tourist experience and a worship experience inside the same building. We had gone through the same doors into a museum one day and the next day a church. Which is the point of this whole discussion. If there is one thing in Britain besides the monarchy that is a winner for tourism, it is the country's cathedrals. In both cases, tradition, history, and the past mingle with the reality of what is very much the present. King Charles might occasionally ride in a gilded coach that looks like a pumpkin, but he also has a constitutional role as head of state under which he must read the throne speech to the parliament of the day. Tourists love the aspects of Britain that are a bit of a theme park of its past. But the delicate balance comes in the places where past and present, showpiece and function, intersect.

There are forty-two Church of England cathedrals in England and Wales (and twenty-four Roman Catholic cathedrals). Built largely between 1000 and 1500, they were constructed in both Norman and Gothic styles—sometimes elements of both within the same building. Winchester Cathedral, in the south, was begun in 1079; Yorkminster, in the northeast, was constructed between 1230 and 1472. All are, without exception, massive, gorgeous buildings. Each is the seat of a bishop and a diocese. Each has its fundamental place in the history of local or national England: the remains of the Venerable Bede are buried in Durham; Hereford, as previously mentioned, is home of the thirteenth-century Mappa Mundi. Westminster Abbey is the burial site for 3,300 people including sixteen monarchs, eight prime ministers, and Charles Dickens. Chichester, going a bit against the theme of ancientness, has some extraordinary works of modern art by such masters as Marc Chagall and John Piper.

The tourism draw of a cathedral is irresistible. When you come upon one, it is impossible not to want to go in. It is like coming upon the Alps or the Rocky Mountains but having a door through which to go inside.

Then, once inside, the immediate sensation is the diminishment of the self. This is so big, we are so small; this is so significant, I therefore am insignificant. But not in a way that I dislike or that terribly discomfits me. The effect from the very conception and building of Gothic and before them Norman cathedrals was to overwhelm simple humans with the majesty of God. However, it's a safe majesty. There is nothing threatening about being inside a great cathedral, even in the vicinity of the crypts with their multitude of bones and collection of dead bodies; the presence of God guarantees the safety of even tiny insignificant little old me. To inspire awe was the premise and purpose of the great churches in the Middle Ages. Religious awe and, to some degree, political awe for the public to acknowledge the powers of both the clergy and the secular benefactors. Worshippers over centuries experienced this. And even an irreligious tourist senses this. That is why we are attracted.

Our other fascination is this: How did they possibly build such confabulations way back in the eleventh century, when we, with all our engineering and technological know-how and apparatus, don't believe we could possibly do so today? Many of Europe's great cathedrals were built within a fairly compact span of time, basically the tenth and eleventh centuries. In England, Lincoln, York, Winchester, St. Paul's, Durham, Hereford, Norwich, Ely, and Wells were all started between 1072 and 1192, with some Johnny-come-latelies like Salisbury and Southwark weighing in twenty or so years later. This was an extraordinary binge of construction for a population of something less than three million people. Not to mention they were not exactly in adjacent neighborhoods, but widely spread out (Durham in the north is 300 miles, or 480 kilometers, from Winchester in the south). And by and large these enormous churches were not to be found in the cities but in the sparsely populated countryside.

The chain of causality for this flurry of cathedral building came from across the English Channel, with the invasion in 1066 of William the Conqueror. William was a Norman, and Normandy was, at the time, the home of both high levels of religiosity and increasingly sophisticated levels of church architecture. After enduring pillage at the hands of invading heathens in the tenth century, virtually all the cathedrals and major abbey churches in Normandy were rebuilt grandly in the course of the eleventh century. So explains Christopher Wilson in his dense book, *The Gothic Cathedral*: "And, after the conquest of England this same comprehensive approach was employed across the Channel on an even larger scale with even more spectacular consequences."[1]

In Normandy, all the construction activity was paid for through the vehicle of lavish ducal and aristocratic patronage of monasticism. For all intents and purposes, William imported this attitude to Britain; bishops, senior clergy, and local dukes were all tapped to produce or raise construction funds. A big fundraiser was the selling of indulgences—the get-out-of-jail-free cards of their day whereby one could, for a donation, buy forgiveness for one's sins. The French influence was apparent not least in the offerings of the Cistercian order of monks, skilled builders who, after their arrival in Britain in 1128, rapidly set up some ninety splendid abbeys.[2]

Even the casual observer will note a pattern in the design of cathedrals, particularly those of a period. These are not quite like the boilerplate blueprints employed by a nationwide chain of fast-food restaurants or Walmart stores, but not entirely different either. There is a basic consistency. In the case of medieval cathedrals, it was that the building conform to the shape of a cross; that its entrance be from the west and that it, on the whole, be oriented east, toward Jerusalem and the rising sun. This morning sun inevitably flooded the building as it beamed in though great windows above the chancel and the altar. The cruciform plan produced both a transept and a nave—the main space of the church where the body of congregants are seated—longer than the other three arms. To the sides were vaults and aisles and arcades. Each cathedral would have one or more towers integrated into the main body of the church, usually over the center of the cross(ing) or on the west facade of the nave.

The influence on church architecture in Normandy, Wilson tells us, was Romanesque and characterized through the device of the "bay":

> In architecture, any division of a building between vertical lines or planes, especially the entire space included between two adjacent supports, thus, the space between two columns, or pilasters, or from pier to

pier in a church, including that part of the vaulting or ceiling between them, is known as a bay. . . . Virtually all Gothic great churches inherited the basic premises of their design from the most highly evolved traditions of Romanesque architecture, those developed in northern France during the 11th Century.[3]

Religious imagery and symbolism was part of the plan, naturally, without question. On the premise that a church was to be an earthly symbol of Heaven, a great cathedral's job was to summon up images of Heaven. If you're ever in the very small County Somerset city of Wells, have a look at the number of niches on that cathedral's west face, and you'll see them filled to almost overflowing with statues of the saints. This is meant to be evocative of "the many mansions of the heavenly Jerusalem" (John 14:2). The smallness of the portals is presumably inspired by the Sermon on the Mount: "Enter ye in at the strait gate" (Matthew 7:13). All such symbolism established the tone of both worship and, latterly, even the tourist experience.

A big deal in the twelfth century were relics. The faithful could be made even more faithful should they be persuaded that the bones of some saint were nearby to their worship. Relics became a driving force for financing the building of a cathedral following the charge led in Britain by Canterbury. There, in the 1170s, the rebuilding of the choir section, adjacent to the altar, got fueled by the St. Thomas Becket story, he who had met his untimely demise inside that very church. With the different sensibilities of our age, relics don't possess quite the same power with most of us. But it inspires a kind of eeriness to be made aware that they are still there.

A cathedral may have been a mind-boggling entity to construct in the first place, but a millennium later it is likewise mind-blowingly expensive to run. Cathedrals have staffs to pay, certainly, but thousand-year-old buildings also need tremendous amounts of upkeep just to keep the roofs on, the windows cleaned, and the walls up. Chichester Cathedral in West Sussex recently (2023) completed a reconstruction of its roof. Five million pounds. The cathedral website provides insight into the project and the challenge:

The Cathedral's medieval timber roof is of historical and architectural significance, representing a key phase of English structural carpentry. The magnificent thirteenth century oak structure survives for the entire length of the Cathedral (with some Victorian interventions in the Transepts following the collapse of the Spire in 1861). The roofs had been leaking for many years, creating a damp environment and causing the ancient timbers to rot, with water penetrating the interior plasterwork. The final, and largest, phase of our major roof restoration is now underway—the Cathedral Nave. The Nave roof is of particular importance, as the medieval markings and pegs in this area give evidence as to how these enormous timbers were assembled, and how the craftsmen would have negotiated access without modern scaffolding.[4]

In 2020, Chichester, which claimed it had lost a million pounds in the previous year, was the beneficiary of a £297,000 grant from the Culture Recovery Fund for Heritage and the Heritage Stimulus, funded by the UK government and administered at arm's length by Historic England and the National Lottery Heritage Fund. The importance of this bailout was likewise explained on the cathedral website: "The Cathedral costs £3,000 a day to maintain. The funds from The National Lottery Heritage Fund and Historic England will contribute towards essential existing job roles and operational activity."[5]

Gloucester Cathedral in Gloucestershire spent £5.6 million on its roof in 2019, including approximately £100,000 for solar panels and associated lead repair, re-guttering, scaffolding, and so on. It too received nearly £4.5 million in National Lottery Heritage Fund grant funding, while other moneys were raised from trusts and private donors. Such substantial public grants, however, are extraordinary in that they tend to be one-timers rather than consistent funding such as might be available in, say, Germany where churches are an official state enterprise (half a billion euros a year). So the cathedrals of England must constantly dance to keep themselves financed, hoping for—but not able to guarantee—sufficient collection plate donations. And the financing needs are huge. Durham Cathedral, in the northeast, recently claimed that it cost £6 a minute to keep it operating. The range of costs according to cathedrals' own estimates vary from £6,000 a day for Ely, £12,000 a day for Salisbury, £18,000 a day for Canterbury, and £20,000 a day for Yorkminster.

Each of England's cathedrals has its story, but each likewise has its modern-day challenges to stay alive. These, particularly the financial challenges, are no small matter. In each instance, what were constructed as churches and places of worship and ecclesiastical activity have had to veer frequently to more-unholy sources of income. Gloucester Cathedral, in the west, was able to rake in a great deal of cash as one of the sets for the Harry Potter movies—eight of them made between 2001 and 2011. But most of these great churches have turned to some level of paid tourism. When you walk through the door of Winchester Cathedral or Hereford or Salisbury or Gloucester or Westminster Abbey, the first thing you will encounter (except perhaps for the queue of persons lining up to get in) will be a ticket window. Recent rates (2023) show that St. Paul's Cathedral in London charges individual walk-up tourists £23 and those coming in groups £20.50. Entering Westminster Abbey will cost £27 and Yorkminster in York, £22—which includes a visit to the tower. Nobody sees this as a problem; the assumption is that coachloads of tourists will expect to pay, and their hosts will want to get something from them.

Durham, which was identified as a UNESCO World Heritage Site in 1986, had 727,367 visitors in 2019. Westminster Abbey in London (without a bishop, no longer a cathedral though it once was) had 1.6 million visitors in 2019. St. Paul's, the iconic domed cathedral designed and built by Christopher Wren in the middle of London, presents itself as an active working church with services each and every day, as well as an extremely popular visitor destination for people from all around the world. Its statement reads:

> At St Paul's we receive virtually no government funding and as such, are fully responsible for raising all the income needed to operate. It costs at least £10 million a year to keep the Cathedral open. Although attending to worship is always free, we are extremely grateful to all those who make a paying visit to the building. Not only do you get to see a beautiful and fascinating building, but the charge you pay is what enables the Cathedral to be open and in good repair for people to visit

as they do. By paying to visit St Paul's, you are not only having a great experience, but you are enabling others to share that experience in the future. You're also making it possible for around 780,000 people to join us in worship every year for which we make no charge.[6]

The less-than-subtle suggestion of "paying" to visit is the clincher here. The reality is that tourists are now an increasingly solid part of almost every cathedral's game plan, and the business plan of that is acknowledged and welcomed. Keeping the cathedrals open through paying tourism is a positive. What is a matter of concern, however, is what comes along with those tourists: the wear and tear on each building of increased numbers, and also the threat of turning Christianity from an active religion into a museum piece. Lots of non-believers would say that's already been accomplished but, no matter, it's a serious question.

The promotions deemed necessary to attract tourists signal some other dangers. Churches throughout the ages have not been strangers to wacky gimmicks, either to get money or get people into their buildings. Back in the Middle Ages, the relics industry, again, was the beginning of this. If you could persuade a credulous public that what you had in the box were the bones of some beloved saint or, even better, one of the apostles, or if you could hold up that sliver of wood as a piece of the "true" cross, you had made a great leap forward toward securing both your funding and your attendance. More recent schemes are of a less mystical nature; they include a mini-putt golf course down the nave of Rochester Cathedral and a gigantic LEGO project inside the cathedral at Chester.

Taking the cake for a brief time, however, was a 2019 enterprise in the cathedral of Norwich in Norfolk, where a fifty-five-foot-high "helter-skelter" was installed inside the main nave space of the church. A helter-skelter is basically a carnival ride composed of a slide that circles down around a central tower. The installation was controversial and became instantly a matter of scorn in the international press. Those who argued in favor of it said it "would give people a different view of the inside of the building." Megan Specia, writing in the *New York Times*, quoted the Reverend Canon Andy Bryant, Norwich Cathedral's canon for mission, as saying "he was inspired to install the carnival slide after visiting the Sistine Chapel and admiring Michelangelo's handiwork on the ceiling. It made him think of Norwich Cathedral's own ceiling, adorned with medieval carvings called bosses that depict scenes from the Bible, and lament the fact that visitors could not see them close up."[7]

Norwich Cathedral, which remained proud of the fact that it charges no entry fee to visitors, did ask £2 ($3.50) for a ride on the slide. The critics who lined up on the opposite side were appalled. The Right Reverend Dr. Gavin Ashenden, one-time chaplain to Queen Elizabeth, accused the clergy at Norwich Cathedral of being "unprofessional" and said they were "making a mistake about what a cathedral is good for." He said there was no evidence that tourists become Christians and "just to put in entertainment is naff [vulgar]." "For such a place, steeped in mystery and marvel, to buy in to sensory pleasure and distraction, is to poison the very medicine it offers the human soul," he told the BBC.[8] It took ten days for the contra side to prevail, and the carnival ride was dismantled.

The nine-hole mini-golf in Rochester Cathedral's nave, however, remained in place. As in Norwich, its defenders did their best to reach for theological arguments to support it. Trading on the feature of the little bridges one has to putt over to chase one's golf ball, the cathedral was quoted as saying, also on the BBC, it hopes visitors might learn about faith and "building both emotional and physical bridges." Reverend Rachel Phillips, canon for mission and growth at Rochester Cathedral, said: "We hope that, while playing adventure golf, visitors will reflect on the bridges that need to be built in their own lives and in our world today." Opponents countered calling the course, which takes up almost the entire space of the nave, as a "really serious mistake that 'tricks' people into a search for God." The opinions of visitors were mixed. A visitor to the cathedral was quoted as saying: "It's really nice for children to come here, be able to see the history of the cathedral and have fun." One boy who played on the course added his two cents worth by saying: "I think it's quite a good place for non-religious people come in to experience what it's all about."[9]

Meanwhile in Chester, just a bit southeast of Liverpool, a well-known BBC presenter made the laying of the two-hundred-thousandth block in the cathedral LEGO project an event for her TV show. *Breakfast on BBC One* presenter Louise Minchin did the honors for the UK's most-watched morning program with over six and a half million viewers. This was October 2019, and there were still one hundred thousand bricks to go to finish the replica of the cathedral. Visitors were paying £1 per brick for the privilege and fun of participating in the construction.

The debate over how to sort out a proper system to welcome, accommodate, and then get money out of tourists has been going on inside the cathedral community—and the Anglican community in Britain—for more than a decade.[10] Many church people remain sheepish about the collection of fees from visitors. The principle at stake is that of free access to the houses of God. People look at scripture and see that Jesus threw the money changers out of the temple in Jerusalem (Matthew 21). On another level, however, the pricing of admission has rarely been an issue. Wealthy people used to buy pews to assure reserved seating; big donors got to be buried in the crypts. In the current climate, the suggested saw-off seems to be: if you come as a worshipper, surely money should be no barrier; a free will offering is the best than can be expected. But if you come as a tourist or as if to a museum (or to play mini golf), then surely an entrance fee is expected. Along with purchases at the inevitable gift shop and cafe.

This road, though, remains a bit rocky. A study in 2017 found that while attractions such as galleries, museums, theme parks, farms, and historic properties across the UK had all enjoyed both a rise in visitors and a rise in income in the previous year, places of worship were an exception: The numbers of visitors to those that charged entrance fees had actually dropped by 1 percent. Was it the entrance fees that were discouraging people? "It is worth noting," the report added, "that places of worship were also the category with the highest increase in admission charges."[11] Things are not completely settled either philosophically or in practical terms, though the connection between paying visitors and the ability to operate the cathedrals is becoming more apparent with each passing year. In 2021, due to the COVID pandemic, visitor numbers at Westminster Abbey in London dropped to 163,000 from its previous 1.5 million. The concomitant drop in income forced the abbey to lay off 20 percent of its staff.

Tourist visitors to cathedrals are of two types. A family of four who've just left their hotel after breakfast and spotted an impressive building might be spontaneous drop-ins. Then there is the planned tour, booked perhaps weeks or months in advance. The latter might even be a pilgrimage of serious religious persons, or a group knowledgeable about architecture or history. Geoffrey Sangwine, an Anglican priest from Toronto with an interest in both church history and liturgy, has accompanied several tours as a kind of assisting director. He describes the participants as people whose interest is informed and serious. A typical trip might take twenty Canadians or Americans to, say, eight south of England cathedrals. A good grouping, says Geoffrey, would be Canterbury (the dean of English churches going back to St. Augustine in the seventh century), Rochester, Chichester, Salisbury, Winchester, Wells, Bath Abbey, and St. Albans. A kind of circle tour can be easily negotiated by bus from one town to the next, one overnight hotel to another, with plenty of time for each day visit. Such a tour group would set off with both an interest in and a knowledge of the churches that the trip could only enrich for them. It would be both a holiday and an experience.

Getting down to the brass tacks of exchange: For each cathedral, there would be exposure and the chance to show itself off, and then income from entry fees, guide fees, and doubtless purchases at the cafes and in the gift shops. For the travel industry in the UK more broadly, these enterprises stand as a substantial brick in the edifice of their economies. Airplane fares are paid, hotels are used, meals are purchased and eaten, coaches are hired. On any one such trip, it's easy to account for $60,000 to $120,000 transferred into the tourism economy. Across Britain, calculations suggest that the spin-off impact of cathedrals on their local economies amounts to something in the order of £91 million per year, with a total broad impact of perhaps £150 million per year.[12]

On the topic of whether they are visiting churches or museums, Geoffrey Sangwine says it has always been the case that cathedrals were a bit of both; they were always repositories for historic and significant artifacts. As a Christian and a clergyperson, Sangwine says he is not offended by churches operating as public spaces: "Churches have also always been that." However, the caution: "At what point do you become a fairground instead of a sacred space?" Worse, possibly, than being a mini-putt course or a LEGO assembly venue would be to reach the place where you feel yourself "pretending" to be what you in fact are in order to satisfy a demanding but unsophisticated audience. Critic Dean MacCannell, in his book *The Ethics of Sightseeing*, talks about what he calls "staged authenticity," which is the impulse to please an audience through exaggerating the elements of what you are in order to perhaps make yourself more accessible or to meet expectations that are less than nuanced. This could mean that for the benefit of non-believer tourists, cathedrals would become sort of "pretend churches." MacCannell warns that "once staged authenticity becomes general it will take a huge ethical shove to put the human back into the social."[13]

Behavior and decorum in a religious space is often a life-or-death matter to the seriously devout. To the tourist visitor, it is more likely to be merely spectacle. I recall one time visiting the Church of the Nativity in Bethlehem, where some of my fellow tourists who were serious pilgrims were near to despair due to having been hustled through the grotto where the birth of Christ had supposedly taken place two thousand years earlier. These were folks who had waited all their lives for this sacred moment, only to be rushed along by mobs who wanted only the chance to take a selfie.

Churches, traditionally, have a lot to do with restricted spaces. In Judaism, only Levites, for example, were allowed to carry the Ark of the Covenant, and even they would drop dead should any one of them actually touch the Holy of Holies itself. In early Christian churches, the narthex, or outer entry hall, was where new converts were obliged to stay, as they were not yet permitted into the church proper. Even today in Catholic or Episcopal churches, worshippers are not really supposed to advance beyond the nave, the choir and certainly the chancel and altar being reserved for those performing the mass. In Christian symbolism, the nave (even though its name has maritime roots in Latin) represents Earth; the chancel, Heaven. So the regular folks stay to the Earth while the elect of the clergy and priesthood operate from Heaven. Worshippers honor these distinctions and divisions, but you have to be careful—unless strictly guided and educated, tourists might roam wherever they please.

The story throughout seems to be about compromise and watchfulness. As in Venice or Barcelona, is there a point where there will be one tourist too many, a moment when the balance tips and what thinks of itself as still an active place of worship becomes purely a museum? Who is keeping an eye on that? Who is worried about that? And, by the same token, is anybody worried that no matter how many tourist visitors you might have and how high the entry fees, it will never be enough to keep the lights on and the roofs repaired in these magnificent though very demanding structures?

CHAPTER 6

Winery Tourism

What It Really Costs

THERE'S SOMETHING ABOUT VINEYARDS THAT MAKES THE HEART BEAT slower. Maybe it's the latent "French thing," fantasies we harbor about languor and elegance in an atmosphere of soft breezes and warm sunshine. Vineyards, even when there is work going on, trimming or harvesting, are generally peaceful, dreamy places. You can hear birdsong. Therefore, there is something natural about our being attracted to them; they are places we are happy to visit, even if we are only pedestrian wine drinkers. And if offered the opportunity to linger, settle into a chaise lounge, and be offered a sample in a sparking glass brought out on a silver tray, we're delighted to indulge. "Wine tasting" is one of those faux snobby activities done sometimes with tongue in cheek but almost irresistible to all but the most churlish.

For these reasons vineyards and wineries, the world over, have become places not just where grapes are grown and harvested and wine is made, but where tourists flock. We go for what is on offer—to buy some bottles directly from the winegrowers. But we also want to linger, to make a day of it, to soak in the atmosphere, or to have a special experience in a special place. The owners of wineries have not been oblivious to this impulse. In response to the public and touristic hunger, they have enlarged the experience, with restaurants, shops, patios, and wine-tasting bars. Some have gone even bigger, adding art galleries, conference centers, or gazebos inside which one can conduct a wedding. Adjacent to all this, they have been required to build parking lots, walkways, and lookout points. The accommodation of visitors has itself become an industry equivalent to the industry of growing grapes and making wine. So it is across the world and even in tiny corners of relatively new vineyards such as one can find in a couple of regions of the Canadian province of Ontario. One of these regions is Niagara.

Everybody in the world is aware of Niagara because of its stupendous waterfall. Niagara Falls is one of the great natural wonders of the world. However, unlike other major waterfalls such as Angel Falls in Venezuela or Victoria Falls in Zimbabwe, it is right smack in the middle of one of the most heavily populated parts of not just the country but the North American continent. The convenience of its location has had a lot to do with Niagara Falls, for 150 years, being a prime tourism destination, a must-see site for both foreigners and also the hundreds of thousands of Canadians and Americans who live within a two-hour drive of the spectacular cascade.

As a tourist site, it sports all the accompanying paraphernalia to amuse the visitor and draw cash from their pockets: wax museums, boat rides, carnival-like arcades, casinos (one on each side of the international border), theaters, revolving restaurants. A walk along a street called Clifton Hill in the town of Niagara Falls can make you dizzy with every kind of imaginable kitsch: arcades, junk food emporia, T-shirt vendors. Brighton in the UK or the US's Atlantic City can't possibly outdo Niagara. The US-based vacation rental and property management company Casago declared the town of Niagara Falls (population 48,000) Canada's biggest tourist trap; it ranks in seventh place among global tourist traps.[1] It has everything required to give it that designation. You can take zip-lines over the falls; line up for boat rides under the falls; go to places where you can watch movies of the falls, if you'd rather do that than observe the real thing. And mobs of tourists do all of the above.

For the economy of tourism, Niagara Falls does its job. The data for visitor numbers and economic impact is tricky to settle on because it comes from two sources, Ontario Parks on the Canadian side and New York State on the US side. But generally agreed-upon numbers settle around thirteen million annual visitors and $3 billion USD in economic activity. This is a massive tourism enterprise. The economic impact on the Ontario side alone is 2,800 businesses and 40,000 jobs.

Tourism, however, has not been alone among the economies of the Niagara region. The tourism industry surrounding the falls has been in competition, first, with another economic driver: hydroelectric power production. In the nineteenth century, the cascade of the waterfall was irresistible to nascent industries. All that falling water could not possibly be let to go to waste; water-driven mills sprang up in abundance, especially on the US side, making upper New York State a localized industrial powerhouse. With the development of electricity, this use of waterpower soon got supplanted by channeling the falling waters into pushing hydroelectric turbines—on both sides of the river and both sides of the border. This activity meant that tourism and heavy industries quickly became competitors.

Over a century and a half, the push and pull between using the hydro capacity of the falls for industry and making a buck out of the visitors who flocked there for tourism intensified before it ended in a saw-off. The competition, according to Daniel Macfarlane in his fascinating book *Fixing Niagara Falls*, came close to completely ruining the natural wonder.[2] Few of us probably realize that more than half of the water that would once have flowed over the falls is even now diverted for other purposes—some of it into shipping canals, the Erie and the Welland, and the rest for power generation. A startling tale recounted by Macfarlane is that there was a time in the early twentieth century when enterprising engineers proposed—and claimed to possess the capacity—to turn the waterfall off entirely whenever they might want to do so. This apparently wasn't as far-fetched as one might think: an idea floated at the time was they could shut the waterfalls down for most of the week in the interests of industry, and just turn it back on for the weekend for the tourists. In retrospect it is doubtless we can all thank tourism for the fact that such a maniacal scheme never achieved fruition.

Power generation and tourism are both immensely important to the life and economy of the Niagara region. However, there has always been a third player too. According to up-to-date calculations, 64 percent of the land in Ontario's Niagara peninsula remains agricultural, which makes agriculture truly significant as well.[3] This Niagara agriculture is of various sorts, but central among it is something arguably more precious than gold. The Niagara Fruit Belt is one of only two areas in all of Canada (the other being British Columbia's Okanagan) where it is possible to have a substantial commercial production of tender fruits and vines—not something you can find just anywhere. The right combinations of soil and microclimate make this possible, but the opportunity is limited. In the case of Niagara, according to an organization dedicated to the preservation of agricultural land, the Preservation of Agricultural Lands Society, of 232,817 acres of farmland in Niagara, only about 35,000 acres, or 15 percent, has the right mix of soil and climate for the precious crops the region has become famous for, tender fruits like peaches, plums, cherries, and grapes.[4]

GRAPES, A HISTORIC CROP

The first European settlers in the Niagara region found there to be already an abundance of wild grapes as well as fruit trees. The climate and the soils, the protection of the Niagara Escarpment—the geological feature that provides a sheltering wall over the region—and the moderating effect of the waters of the lake made things kind of Garden of Eden perfect. The settlers picked up on this and got to work; as early as 1811, a local entrepreneur bottled small batches of commercial wine. By the 1850s, large plantings of grapes had been undertaken. It turned out at the time that only the domestic ones thrived, as imported European vines suffered from fungus and could not take hold. Purple Concord grapes were the most hardy local variety and, a century later, by 1980, they continued to represent 27 percent of the crop.[5]

A great time for Ontario vintages was the era of Prohibition, from 1916 to 1927, during which the provincial Temperance Act made domestic wine from Ontario-grown grapes the only alcoholic beverage that was legal. Niagara grape growers prospered. Thirty-three new wineries were established with, by the 1940s, one big company, Bright's, being the largest producer. By then some success was had with French hybrid vines and, three decades on, in 1971, 76,380 tons of grapes were produced.

FRUIT, LIKEWISE A BIG CROP

What also thrived in the sandy soils were tender fruits. Tender fruits is another name for cold-sensitive fruits such as peaches, apricots, sweet and sour cherries, plums, and pears. Among the earliest to plant such orchards in Niagara were refugees from the French Revolution who imported a variety of fruit trees from their homeland. Over the first century of European settlement, the growing of both grapes and tender fruits became an industry with a ready market among the whole of southern Ontario's grocery stores. The peaches, plums, and grapes that Ontarians consumed came almost entirely from Niagara, which made the region an invaluable garden supplying the tables of thousands of homes.

Nothing, however, is ever completely secure. The precariousness of fruit production in the Niagara region became a flagged issue as long ago as the 1950s. "The Disappearing Niagara Fruit Belt" was the title of an article in the *Canadian Geographical Journal* in April of 1959.[6] In 1980, a graduate thesis for McMaster University examined the same topic.[7] A number of factors impinged: The already confined land base was up against the relentless push of urbanization and industrialization. Driving around the peninsula today, one is alert to a host of pressures on this relatively confined territory. You readily spot vast tracts cleared for new housing developments as the pressure of the growing population spilling out from the city of Toronto—albeit two hours away—exerts its influence. As well, there was emerging an internal competition between different forms of agriculture. While 64 percent of the land in the Niagara peninsula is agricultural, in the last three decades, a competition between different forms of it has intensified.[8]

The effects of these competitive pressures have been more obvious on the fruit trees than on the vines. Thirty years ago, a book titled *Niagara's Changing Landscape* pointed out that, sadly, while "such was not the case with the wine industry," the tender fruit industry seemed to be in decline.[9] More was to come. The wine industry, for its part, was, at that very moment, about to undergo changes that would lead not just to its survival, but to its explosion. Until about 1976, when the United States stopped importing them due to competition with vineyards in New York

State and Michigan, Concord grapes for juice remained a big export product out of Ontario's Niagara region.

A decade and a half later, in 1990, with international trade rule changes from NAFTA and GATT, the grape business in its entirety went topsy-turvy. These two trade regimes, the North American Free Trade Agreement and the global General Agreement on Tariffs and Trade, altered all kinds of rules, not least among them those around wine exporting and importing. The upshot heralded almost instantly the emergence of Niagara as a producer of higher-quality wines than had hitherto been the case. Native "labrusca" grapes, among them the Concords used for jellies and juice, were largely pulled out while newer French hybrids and vinifera, or wine grapes, were planted. Some of these were ideal for the making of ice wines—where the grapes are picked only after being touched by frost. This charmingly sweet dessert wine would quickly become a famous staple of the region.

What was about to occur was the rollover of one predominant crop to the next and also the arrival of the right to use an agricultural metaphor for something not naturally agricultural. The crops of tender fruits turned into crops of grapes that turned into crops of tourists. More and more wineries began to set up, and many of these quickly realized the potential of not just making wine but aligning themselves with tourist and cultural promotions, a kind of tourism that got its own name, "agritourism." From there, things only exploded: on-site shops, wine-tasting venues, restaurants, conference centers, and parking lots all plopped into place.

The competition for limited land likewise heated up. With not "agri" but "tourism" as the main factor, land for grapes became more valuable than land for orchards. Orchards got torn up, wine grapes got planted. Inside the blunt reckoning of capitalist economics, this was all defensible. As a fellow who runs a large wholesale produce business out of Toronto, Vince Carpino, when asked about the hierarchy, told me how a peach sells for a dollar, a glass of wine for five dollars. To Vince, what had happened was a no-brainer. Besides, peaches for Ontario's dinner tables could now come, and come in abundance, from faraway California and Mexico. Households in Toronto could now get their peaches from those sources instead of from orchards close to home—and have them available pretty much year-round.

But where did this leave the "agri" part of agritourism? Gracia Janes has been trying to prevent the disappearance of agricultural land on the Niagara Peninsula since 1974 when she became a founding member of PALS, the Preservation of Agricultural Land Society, headquartered in her hometown of Niagara on the Lake. Though it has not been entirely a losing battle, neither has the war been won.

To Janes, the fruit land is a precious and very limited commodity that has been shrinking. She points out that in 1974, the Niagara area had 35,000 acres of tender fruits and 35,000 acres of grape vines while, at present, the fruit acreage has been cut by more than two-thirds. Fruit acreage dropped down to 10,000 acres. A good part of this loss has been due to the increasing sprawl of urbanization. The PALS organization along with a number of other community and industry organizations have tried—and come close a number of times with successive provincial governments—to enact "restrictive covenant" programs that would keep all the valuable fruit land permanently protected. To date, this is still a hope rather than a reality. It makes her sad, she told me, how every time she drives down Highway 55 (the road coming off the Queen Elizabeth Freeway at St. Catherine's, the main highway from Toronto, and proceeding into Niagara on the Lake) she is confronted with all the new and pending development. But there is also no question in Gracia Janes's mind—agreeing with Vince Carpino—that a certain proportion of the decline of the tender fruit (peaches, cherries) acreage is due to the expansion of vineyards and, of course, the wineries and "other tourism" that goes with them. "Highway 55 is all vines now," she says.

On top of that come the spin-offs that are exceedingly costly to the preservation of fertile lands. Many new industrial initiatives, Gracia Janes observes, have to do with the wine industry, though not its agricultural part. Increasingly large wineries, and what she refers to as "the inter-ference of tourism uses" such as restaurants and tasting venues and the parking lots that go with them, are taking up more and more precious development spaces.[10] Winery-based tourism is a recent, but now a solid, part of both the culture and the economy of the Niagara region. Figures from 2023 state that the wineries sector in Ontario was responsible for $5.49 billion CAD in the provincial economy and 22,385 jobs. The tourism component of wineries made $1.04 billion and created 5,888 jobs; 2.6 million tourists visited Ontario wineries.[11]

It is pretty much too late to ask whether this is good, or whether it is sustainable. The shift to wine and wine-tourism seems a done deal in what forms the current story in Niagara. With two significant results: Yes, the region does produce some very nice wine, but the main raison d'être for the plethora of wineries that have sprung up is not the wine they produce but the tourism trade they entice. Both the Niagara region and Prince Edward County, which we will look at next, were once prime fruit and vegetable growing regions whose farms and orchards supplied produce for far-flung populations. Both regions have lately been taken over by wineries for whom, arguably, the production of wine is almost a secondary purpose. The main purpose is to attract tourist visitors and host events such as conferences or weddings.

PRINCE EDWARD COUNTY: ANOTHER REGION, SIMILAR PRESSURES

Another spot experiencing the mixed blessings of winery tourism is Prince Edward County. Unlike Niagara, vineyards do not appear to be the villain squeezing out crops of tender fruits there, but viticulture has still become a prime factor in the local economy with plenty or adjacent spin-offs, some positive, some negative.

Settled in the 1700s by United Empire Loyalists, refugees from the new United States who had remained attached to the British Crown during the American Revolution, this region, a 1,000-square-kilometer (400-square-mile) peninsula jutting into the north side of Lake Ontario about 200 kilometers (120 miles) east of Toronto, was originally forest that then got turned into farmland. The county's agricultural history includes a lot of general farming with livestock and crops but also, importantly, the extensive growing of fruits and vegetables. Due to just the right kind of soil and a lake-moderated climate, apples, cherries, tomatoes, corn, and beans became predominant crops in the area for almost two centuries.

A dozen canneries were set up to process the harvest and form the foundation to distribute the product far and wide. Ontario's first cannery was opened in the town of Picton in 1870.[12] In 1882, native son Wellington Boulter, later to be known as the "father of commercial canning in Canada," opened his cannery, also in Picton, to handle locally grown corn and tomatoes. Over the next decades a plethora of plants opened in nearby Brighton, Bloomfield, and Wellington to process local crops of tomatoes, corn, apples, peas, and raspberries. Jams and jellies, catsup (ketchup), and chili sauce were all made as well and transported by rail to the population centers of southern Ontario as well as to ports for export to the United Kingdom and Europe. The heyday of all this lasted about a century. By the 1980s, however, a combination of changes in transportation, alterations to the rules of international trade, and the development of new technologies to freeze rather than can produce caused everything to fall in on itself. The Prince Edward County lands that had grown fruits and vegetables largely reverted to forage crops for animals and other cash crops. The arrival of vines and wineries did not make this happen, but came in the 1990s on the heels of the changes.

The first grape vines were planted in 1999, on the basis of an educated guess that the calcareous limestone soil that was similar to the soils of the Burgundy region of France might, coupled with the lake-moderated climate, prove hospitable to grapes. The guess proved prescient. By the early 2000s, eight hundred acres of Prince Edward County was under vines, and the region got awarded a formal viticulture designation. It seemed only right that tourism should follow, the next iteration of a cash crop. Then the next logical expansion happened: If one drank, one should also eat, so restaurants appeared. Aggressive marketing kicked in to turn what had been the "Produce Capital of Canada" into the "Gastronomic Capital." Articles in the media started to appear touting the county as a "best-kept secret." It didn't hurt that the region was within only a couple hours of sizable urban hubs: Toronto to the west, and Ottawa and Montreal in the other direction. American visitors could cross the border at Alexandria Bay at the top of New York State just east of the Ontario city of Kingston.

The tourists who frequented the county experienced things other than wineries. There were good beaches, natural sights like birds, and a nascent artistic culture. But the wineries, according to local resident Debra Marshall, provided the magic touch. In 2023, there were thirty-six of them in Prince Edward County, some straightforward vineyards, many of them estate wineries; some of them small and family-run, others larger and more corporate. Many of the wineries included tasting rooms, at least one of them combining an on-site art gallery. Among those that also had restaurants, some had facilities capacious enough to host events such as weddings. Corporately, they linked up to form the Prince Edward County Wine Tour, an approximately 160-kilometer circular drive around which you can stop pretty much every ten minutes or so for a glass of vino should you wish. My wife and I started by visiting one that sported a kind of faux-château restaurant building and another with the art gallery.

In short order, something else happened: Not only tourists came, but people well enough off to want weekend and holiday properties started acquiring land and building increasingly fancy homes either for getaways or to retire in. Living near vineyards has its attractions. Debra Marshall, whose family has lived in Prince Edward County since the 1790s and who herself has both grown grapes and served on local business development boards, is one of the region's thirty thousand permanent residents. She, along with many of her compatriots, has concerns about the directions things are going. Debra points out that there have always been tourists, certainly as long as she can remember—every spring, Americans came to fish and would either camp or stay in local hotels. But with the wineries, the demographics of the tourists have changed. They are more demanding; they have more money. This is echoed by another resident, Donald Walker, who, with his partner Frankie Ip, took up residence at the west end of the county a decade ago. In that time, he says, he has watched not just the local economy but the local demographics get transformed. Donald doesn't require much prompting to provide a list of the changes he has observed in his near-neighborhood:

- Minor beaches that once were popular with local residents now are posted as no-stopping/no-parking zones with $500 fines because of overcrowding, traffic congestion, and neighborhood disturbance.

- A substantial proportion of the houses in Wellington (his local town) have been converted to short-term rentals, driving up housing costs for local residents.

- Restaurants, wineries, and tourist accommodations are unable to attract staff to meet their needs because housing is too expensive.

- One operator offering limousine winery tours had tripled his fleet size.

- The entertainment value of winery tours exceeds the "taste" value of county wine while the market-garden farms and canning industries that filled shiploads of food destined to Britain during two world wars have disappeared along with the harbors where the boats once were loaded.

- Improving internet speed enables work-from-home professionals to enjoy the quiet life by the lake while still being a relatively short commute for occasional on-demand appearances at the city office, and real estate prices continue to increase disproportionately to inflation.[13]

Prince Edward County, getting back to Debra Marshall, is dealing with the same matters so many other places are grappling with. Though not reaching the pushback levels of some of Europe's iconic cities, there is an unsettled mood among the locals. Marshall fears that the rush of tourism possibly got ahead of itself. "We have the goose that laid the golden egg, but we're paving over paradise," she says, not shy about mixing metaphors. In short, the county wasn't ready for the onslaught that came when its promotion of tourism took hold. Touted in marketing as a "gastronomic capital," it got prize-winning chefs for restaurants the locals couldn't afford to patronize. Airbnbs spread like wildfire, but were largely uncontrolled by regulations. Consequently the rents for locals wanting year-round housing went sky high. The locals also became irritated by other economic elements. They weren't hired in the tourism restaurants or winery businesses because they "didn't have the skills." Outsiders, she complains, were imported for many of the jobs and they took up housing spaces. Locals questioned whether the tourist industries were paying a fair share toward the municipal budgets which were all property-tax based or that the priorities of the longtime taxpayers were being met. They grumped that the county needed to "stop spending half a million on marketing and start fixing the roads."[14]

"It's insane" is the way my hostess at a B&B in Picton puts it when asked to compare current tourism demands with those of twenty-seven years ago when she first opened her home to be a bed-and-breakfast. In the next town of Bloomfield, a charming eighteenth-century place originally settled by American Quakers, almost every second stately house along the main street advertises as an inn or bed-and-breakfast, but our hostess Cyndi believes there is no way to properly estimate how many are truly available since she believes that probably lots more remain unregistered. A kind of Wild West attitude has taken over the tourism industry as the county becomes a more and more popular destination.

In the local economy, tourism had taken over from agriculture, which held the top spot for a number of generations. Everything having to do with tourism has expanded dramatically in the last two decades. Accommodations, knickknack shops. Not only people who make wine, but people who make cheese, or cider, or pottery all depend on the tourists as well as being the argument for tourists to come. It needs to be noted that for the most part, tourism in Prince Edward County is seasonal, which also causes whiplash among the locals. In the winter the restaurants are closed, the Airbnbs are empty, the hotels drop their rates, the seasonal workers depart. But in the summer season, the locals claim they can't use the overcrowded local beaches or negotiate a parking space in town to do their shopping.

It would be hasty to characterize all this as an unmitigated negative. It is simply a reality, one that, though unsettling to some, is a natural by-product of tourism and especially its rapid growth. Choices are made. Choices have consequences. In both Ontario's Niagara and Prince Edward County regions, tourism has been a recent growth industry and gives all appearances of continuing to expand and grow. This means change. Wineries are a big part of that change. But the trade-offs of other forms of agriculture in favor of grapes and the role of grapes in the growth of tourism are both realities that have landed on the locals perhaps a bit too suddenly.

What Does Sustainable Tourism Look Like?

BROADLY SPEAKING, THE WORLD OF TOURISM HAS FOUR MAIN GROUPS of stakeholders or players, each of them and their interests critical to the long-term game plan, the sustained health of worldwide travel and tourism.

The first group of these critical players is made up of the tourists themselves. In the developed world, travel for pleasure is the third-largest household expense after food and housing. Those of us who live in that world increasingly believe that our right to travel is sacred. We undertake it with zeal, and we welcome the increases both in our disposable income that we can put toward travel and in the expanding capacity of the travel industry to accommodate us. We take for granted the fact that we can go anywhere in the world within twenty-four hours, and we believe in our right to act on that ability. We do not want limits placed on our right or our potential to travel, to go anywhere we wish. We want borders to be open, we want our travels to be safe, we want to be served and treated well upon arrival, and we hope our traveling can be ever less and less expensive. We are important because it is our motivation that moves it, and it is our pocketbooks that pay the bill. Without us, none of it would happen.

The second group of important players are those who comprise the industry that has been built around tourism. This is the business: all the hotels and Airbnb services, the tour companies, the agencies, the airlines, and the financiers. Many of these players are huge multinational conglomerates; many of them intersect and overlap. The industry is, by and large, knowledgeable and efficient. But the most important thing to remember is that it is a commercial industry, a money-making industry. This is capitalism. The industry that surrounds tourism is competitive and profit oriented; it will do whatever it needs to in order to stay afloat. It is exploitive when it needs to be, and cost cutting when it needs to be; the bottom line is ever present and constantly worried over. Capitalism, put simplistically as we might remember from our basic university economics courses, needs profits and flourishes on growth. The tourism industry growing in leaps and bounds with, until now, almost endless frontiers has been a near perfect example for it. The momentum of the industry, consequently, is huge. It is virtually unstoppable in its movement, and its direction is not easy to shift.

The third group with a substantial interest are the world's countries and governments, the national economies that court tourism as an important component in their industrial strategies. These are not just small Caribbean islands for whom tourism is their monoculture crop and their main foreign exchange earner—all established Western countries have likewise become increasingly aware of and jealous of the role tourism plays in their economies. Tourism in France, for example, represents 8 percent of the county's GDP. Ninety-million foreign visitors came to France in 2019. In countries such as Kenya and Cuba, tourism is commonly recognized as the main ticket for foreign exchange earnings; if these nations want to have money for any other sort of trade, they have to keep tourists coming and tourism healthy. But in almost all other places, it plays a critical role in the national employment. Tourism in Britain is worth more than 106 billion pounds and supports 2.6 million jobs.[1] In Canada, tourists spent $100 billion CAD in 2019 and supported 1.9 million jobs, or 10 percent of the workforce.[2] In Italy it was worth 58.3 billion euros in 2018. In Greece tourism represents 10 percent of employment and 382,000 people (2018). In Israel it accounts for 3.6 percent of total employment, or 140,000 jobs.[3] All over the world, tourism is a sizable player in nations' economies and thus wields enormous clout and leverage.

The last remaining players with a legitimate interest are the visited, the places and people that are the destinations of tourism. This includes the island beaches, the European cathedrals, the ancient temple ruins, and all the people who live near them who are being asked to share with visitors the temples and cathedrals and beaches and wild animals that have been part of their lives for centuries. Thai or Kenyan or Guatemalan villagers and the environments they inhabit must put up with tourism and tourists. They are not necessarily the direct economic beneficiaries of this tourism, yet they have no choice but to live with its onslaught and in its wake. These, of all the players of tourism, have the smallest voice; they usually must struggle for attention even from their own government representatives.

To find the balance among the complexities of these competing interests is tourism's greatest challenge, a challenge about which it has become more and more conscious. This might be considered tourism's bottom line. Thus far, the four interest areas remain out of balance, the four players not nearly equal to one another in strength. But as the voices of each become more articulate and more assertive, finding an acceptable, useful, happy, productive equilibrium will be the test of tourism's, and perhaps even the world's, ability to survive.

THE TOURISTS

It has been observed that while global travel from south to north is mainly a function of immigration, the primary travelers in the other direction, from north to south, are tourists. Not that all tourism goes from north to south—there is plenty of north-to-north tourism with visitors flocking to the cities of Europe or the scenic spots of Japan, Asia, or North America. But the majority of tourist visitors, in all cases, are we from the Northern Hemisphere who are termed "Westerners"—plus Asians. It is our hegemony, at least as of the moment.

What is important to this group that will be decisive in the sustain-ability of tourism? Despite movements within the industry that market, as pointed out in our first chapter, predictability and safety, a substantial hunger of the Western tourist is for "authenticity." This is according to observers like Professor Dean MacCannell. When we travel a great distance, what we want to see when we get there is "the real thing," whatever that might be. If we want Disneyland or Las Vegas, we will go to Disneyland or Las Vegas. However, we don't want to get Disneyland when we are not expecting it. If we want to visit a rainforest, we want a real rainforest when we get there. If we want Roman ruins, they must be authentic Roman ruins. Certain emerging varieties of tourism are built solidly around this factor of authenticity. For example, do-good, or volunteer, tourism where one might travel to a distant location in order to roll up one's sleeves and participate in the building of a school or digging a well fits into this. The visitors feel they are working to earn their keep, getting to know the "real" local people, leaving something useful behind when they depart. How this is seen by the locals is possibly an entirely different matter, but for the tourist-visitor it is, in general, substantially satisfying.

For increasing numbers of travelers, such options fit the bill. But in the search for authenticity there is the temptation to push the envelope. A novel wrinkle in Nairobi, Kenya, is to take visiting tourists on sightseeing trips into the shantytown slums. The city of Nairobi (population 4.4 million), not unlike other African cities, has sprawling areas of informal settlement called "shantytowns." I know these from many visits to medical clinics situated in shantytowns during projects documenting HIV/AIDS research. The shantytown communities are seemingly endless vistas of cheek-by-jowl huts slapped together from whatever is available, generally sticks, corrugated iron, and cardboard. They sprawl: Majengo, the one I know best, houses half a million people. Kibera, the largest Nairobi shantytown, has 800,000 people. They are "unofficial" in the sense that municipal governance and municipal services are minimal. Yet all manner of life goes on: joy and sorrow, labor and love, hardship and cooperativeness. Small boys can be seen at play in stinking open sewers; chickens, dogs, and goats wander everywhere; noisy, open-air markets of food, produce, used clothing, and shoes do brisk businesses for either cash or barter; children in spotless school uniforms emerge from claustrophobic mud huts. Everything is simultaneously predictable—as in what you would expect amid extreme poverty—and astonishing. What these have not been—until recently—are tourist sites. But if you search out something called viator.com, you are offered this:

> Get a glimpse of life in Africa's largest urban slum on a private walking tour to Kibera, a deeply impoverished area that's home to thousands of people. Visit the area with a guide who was born and raised in Kibera, tour an orphanage and school, then see how local craftspeople repurpose bones in a bead factory. This tour includes pickup and drop-off at your hotel in Nairobi.[4]

Justified as "educational," such tours are either the height of authenticity in tourism or the worst sort of voyeurism. For myself, I must say that while entering the shantytowns as a documentary maker accompanying health workers was something I found to be endlessly fascinating and energizing, going into the same neighborhoods as a tourist who has booked a guided walk would, frankly, give me the creeps. What is the difference? In both cases I am the same person, a visiting outsider who has a comfortable hotel room to return to at the end of the afternoon. What it boils down to, I believe, is not the question of authenticity, but of reciprocity. In the former iteration I was able to convince myself, at least, that I was part of a system that was providing something everyday and useful to the masses of people among whom I was walking: care for their health. In the latter, it would be hard to persuade myself I was anything but a voyeur seeking entertainment. What am I offering in exchange? The fee I might pay to the walking-tour guide would hardly land in the pockets of the thousands of people into whose lives I was intruding. What can I possibly give them unless, perhaps, a chance for a bit of a laugh at the absurdity of a mob of obvious outsiders—possibly clad in full safari gear—coming to gawk at them? Hence, the thoughtful tourist must always be pushed to ponder the matter of ethics. And, as MacCannell puts it, the questions to be asked are simple and straightforward:

> Does my presence in this strange land help or harm the people or the natural systems I encounter here? Are they joyful about my presence, indifferent or hostile? Do they seek cynically to manipulate me? And even if their welcome is genuine, does my presence ultimately harm

them, their progeny, or the land in ways neither they nor I can yet discern?[5]

THE INDUSTRY

Industry sustainability in tourism seems to be straightforward. It depends on satisfying the first stakeholder group, the tourist customers, and doing a better job in addressing the growing demands from both the global public and policymakers to improve environmental performance. Everything, from the environmental costs of air travel to waste disposal and water usage within tourism facilities, be they hotels or cruise ships, has come to be under intensified scrutiny. To flourish, the industry will have to be both responsive and creative. It will have to do that while likewise accommodating a continuously growing appetite of more people to travel. The industry is not likely to shrink, but it will have to show that it is both responsible and responsive.

COUNTRIES AND GOVERNMENTS

For the last at least half-century, tourism has been promoted almost universally by governments big and small, national and local, using tax dollars and publicly funded agencies operating on the premise that this "soft industry" would be good for everybody. With few exceptions—some of which we itemize in this book—this remains the operating credo of many jurisdictions. Expanded tourism is good for development, it is good for local jobs, it is good for foreign exchange dollars, it is good for the tax base. Country after country and region after region publish the statistics of their tourism economies. The French tell us that tourism is considered as well to be a "soft power asset" in their foreign affairs arsenal, meaning that as foreigners visit France they come to understand and become sympathetic to France. France has been very good at this and has reaped the benefits of being the world's leading tourist destination for years. A record ninety million international tourists visited in 2019.[6]

Marketing strategies and government subsidies to industry infrastructures remain de rigueur almost across the globe. Governments, almost everywhere, continue to see supports for tourism as an industrial or quasi-industrial strategy, though perhaps without asking the nuanced question raised by observers such as Martin Mowforth and Ian Munt. According to Mowforth and Munt, we should never forget that an economic enterprise is always an economic enterprise. Ergo: "Are the ever-expanding tourist markets and the new, responsible forms of tourism in particular, a smoke-free, socio-culturally sensitive form of human industrialization?"[7]

From the standpoint of governments, however, one thing that *is* becoming increasingly evident is the necessity to respond to their citizens who question whether it is all worth it—a worthy expense for taxpayers to support. They might need to work harder to persuade their citizenry that this whole business is good for them. Or if it be proven that it is not, to be ready to accommodate the necessary changes. Sustainable—as we saw for Venice, Rome, and Barcelona—also means keeping the locals happy.

THE VISITED

The fourth stakeholder group is the one that too often gets left out of considerations: the visited. Statistics show the numbers of people employed either directly or indirectly in tourism and hospitality businesses. And it is a substantial number. But what about all the citizens of a city or country who do not have tourism jobs but, like those we noted in Venice and Rome and Barcelona, have to put up with crowds on their streets, high prices in their cafes, high rents or scarce housing? What about the people who see their once-pristine beaches crowded, their trash cans full, complete strangers asking them to hold still so they can take a photo because the stranger considers the local just going about her business to be picturesque? Once while walking down a back street in Tanzania and blithely snapping away with my Nikon, a woman working in her garden threw a tomato at me. At first I was shocked and offended. But then I had to think: good for her. My presumption and arrogance needed a comeuppance. I was in her world, but was treating her as if she were merely a prop in my world.

From time to time, the effects of tourism on the locals go way beyond mild or benign. We should include in this both local people and the local landscape. A quintessential story comes out of Thailand. In 1999, a popular movie got filmed in a beautiful Thai cove called Maya Bay. Maya Bay is situated in Hat Noppharat Thara-Mu Ko Phi Phi National Park and, up until then, was frequented mainly by locals and handfuls of intrepid backpacker tourists. "The Beach," as it was presented in the movie, of course, represented paradise; it was a beautiful, serene, secret place visited by the hero, Leonardo DiCaprio. The worst way to keep a secret, however, is to feature it in a Hollywood movie starring Leo DiCaprio. In short order after the film's release in 2000, this paradise became instantly and internationally popular, so popular that tourist visitors increased five- to tenfold almost instantly. Their numbers and behavior accomplished the worst: Within a half decade, the coral that had reefed the beach had been destroyed, largely by boat traffic (sometimes there were a hundred boats to be seen at a time), though incoming tourists reportedly walked on the reef as well. The tourists were somewhat "innocents"; most were possibly blissfully unaware of the damage they were doing. But eventually, the beach was left a mess.

Thailand, which by then operated on the understanding that about 20 percent of its GDP was linked to tourism, found itself in a sad bind. Maya Bay had been one of the precious natural resources that created every outsider's conception of Thailand, one of the foundation blocks of its tourism economy. But where was it now? The cove lay ruined; the tourists had disappeared. The locals, who had once had the beach to themselves, now had it to themselves again, but had as well the thankless task of having to rebuild the reefs. This they did over a period of three years. After formally closing the beach in 2018, marine experts and volunteers had to replant over thirty thousand pieces of coral, much of it grown off a nearby island. What will happen next? Noting that the environment of local shark populations had also been disrupted, marine biologist and professor Dr. Thon Thamrongnawasawat told CNN that "the best solution is nobody comes."[8] Whether Thailand's economy and its reliance on tourism can abide that remains the question. There is no small amount of irony embedded in this story in that the film's star, Leonardo DiCaprio, played a backpacker who sought "connection" through paradise.

Sustainability is built on fairness. It is also built around that old-fashioned term that used to be applied to occupations like farming, *husbanding*, as in "husbanding the land" or "managing thriftily," as the *Oxford English Dictionary* puts it. This was manifested through a combination of understanding what you have in hand, taking a modest benefit from it, and being careful not to destroy it for future use.

A woman I talked to in the wine country of Ontario employed a mixed metaphor when she looked at the challenges her community faced in managing its tourist resource. The goose has laid us a golden egg, she said, referring to the attractiveness of the local landscape and the richness of the soils for the crops—including vines they could produce. "But we are paving over paradise," meaning their absence of controls and vision and fairness to all the players threatened to lose it all. In order to avoid that happening, "husbanding" of the resource will have to be employed.

* * *

There is an increasingly substantive academic, sociological, and policy-based literature on the subject of tourism and, particularly, questions of sustainability. A consistently expanding series turned out by Britain's Routledge Publishing commenced in 1998 with the first edition of *Tourism and Sustainability*, a dense, 420-page compendium of questions and issues related to tourism and globalization. The Routledge "tourism handbook" series boasts a dozen titles covering such topics as community-based tourism, tourism in cities, inclusion, pro-poor, fair trade, and so on. The tomes are dense and exhaustive reading for the person on the street (or in the airport departure lounge), but they demonstrate that a great deal of thought and study is being put into tourism at the academic and policy levels.

Another source for statistical data and impact prognoses is the World Travel & Tourism Council (WTTC), a global nonprofit whose members come predominantly from the tourism industry. It is, admittedly, an industry organization whose purpose and interests lie in the realm of keeping the industry going, but its research reaches into many of the questions of broad-based sustainability. Its statement of purpose tells us: "For over 30 years the WTTC has conducted research on the economic impact of Travel & Tourism in 185 countries and issues such as overcrowding, taxation, policy-making, and many others to raise awareness of the importance of the Travel & Tourism sector as one of the world's largest economic sectors."[9]

One thing everyone, all four stakeholder groups, is concerned about is something called "leakage." Leakage is a negative term, the simplest explanation of it being that it identifies how the money spent by a tourist in any given location doesn't all stay there. Some of this is obvious and logical: you might stay in a hotel in Costa Rica that is part of a chain owned by a corporation headquartered in the United States. Therefore, a certain percentage of the money you pay for your room heads back to the corporate headquarters and the company's shareholders. There are many other examples. Globalized supply chains, of course, mean that nearly everything is hooked to somewhere else in some fashion. But tourism, as a global industry operated in such a large way by globalized players, runs the likelihood of more of this than some other sectors.

We were told, for example, that the milk we used at our Cuban resort came as milk powder from New Zealand. Ergo, the portion of our hotel fees that paid for the cream in our coffee went to farmers in New Zealand, not to a dairy in Cuba. This may have a reasonable explanation: The Cuban dairies may have had their hands full supplying milk to the Cuban people, with no excess left over for the tourists. But this being the case contradicts the notion of tourism being a driver within local economies, providing jobs and benefit not just to the frontline workers but to all the ancillary suppliers. (Fruit and fish at our Cuban resort, by contrast, appeared to have been sourced locally.)

Some more jarring examples of leakage are recorded by tourists who purchase knickknacks in gift shops only to discover that they are not locally made handicrafts, but facsimiles of objects that might have been local but have been, in fact, mass-produced somewhere else far away. Leakage is complex, though traceable. Locations frequently find it dispiriting when they discover what a small portion of a tourist dollar actually remains in the host community.

Tourism is huge both as an industry and as a human activity. But for it to be sustained means it has to be sustainable. And to be sustainable, it needs to be mindful of at least all these issues and the interests of all its stakeholders. There is no other way for all of us who enjoy the benefits of being tourists to equitably continue enjoying our pleasures.

CHAPTER 8

Being Maasai—and Engaging Tourism

NEARLY THIRTY YEARS AGO, I WAITED IN A RESTAURANT IN NAIROBI, Kenya, to meet three men. The restaurant was an out-of-the-way spot, some distance from the noise and bright lights of downtown, but it been chosen by the men who had communicated to me through a couple of intermediaries. The room was dark, and very warm; the cooking smells were traditional African, steamy maize meal and braised goat meat. The tables were covered with oilcloth and crowded by people whose conversations went on in tribal languages more so than in English. All this seemed appropriate. The men I was set to meet were elders of the Maasai tribe.

When they arrived, Stephen Ole Senteu, Mark Ole Karbolo, and Charles Ole Sonkol were not as "elder" as I had been expecting. Mark was maybe forty, and the other two still in their thirties. The other thing I remember noting was how they were dressed. In this part of Africa, and certainly at that time, one regularly saw Maasai men and women in traditional garb. The Maasai are tall and in traditional garb their bodies, slender as willows, are wrapped in bright, red cloth blankets; their heads are closely shaved. Heavy bracelets weigh down bare arms, and shoulders are encircled by necklaces of endless loops of colored beads. The most startling aspect of Maasai getup, as we know from decades of *National Geographic* photographs, is the work that has been done to their ears: the lobes are first pierced when they are quite young children, and then they are stretched, on both men and women, into huge dangling loops in order to accommodate elaborate rings of heavy, colorful, uncomfortable-looking beads.

Traditional people all wore handmade ornaments in those days, but Maasai in contact with the modern, urban world were becoming known to accessorize all manner of additional materials into their decoration. I had seen one elderly fellow who had strung bright, yellow empty Kodak film canisters—people were still using film then—on his earlobe; a friend reported seeing a person who had incorporated pork-and-beans cans.

But my three guests, though they had come from a distant village in the country, were in smart, dark business suits. There is a moment of double take when you encounter men whose earlobes are pierced and stretched, extended open loops flapping like spinnakers, who also wear business suits. But these three were there because they meant business—both with me, to whom they wanted to tell their story, and with government ministers to whom they had made a special plea. We shook hands and ordered a tray of Coca-Colas and plates of maize-meal ugali.

I thought about this long-ago meeting recently when I engineered an up-to-date investigation on tourism in East Africa. The three Maasai in Nairobi had explained how far they were from home. Their particular corner of the land was known as the "Loita," a forested area of one thousand square miles inhabited by villages containing twenty thousand of their people on the east edge of the Masai Mara. At the time, they told me, the Loita was the last refuge of traditional Maasai life. The world there was still a place where cattle and trees operated as the centers of spiritual and cultural as well as economic existence. When you greet a traditional Maasai, courtesy demanded your asking first about his cattle before inquiring about his family. "Our forest is the only thing that is of value to us," the men explained. "We graze our cattle in the forest and have our *bomas*, or homesteads, there. We have our cultural ceremonies and rituals in the forest."

The Loita forest is mountain forest studded with hundred-foot-tall cedar trees, eighteen and twenty feet around. The Loita Maasai are not sedentary, but had always been nomads. In the dry season, they and their cattle shared the forest with elephants, buffalo, leopards, colobus monkeys, and hundreds of species of birds, reptiles, and insects. In the rainy season they move, with their cattle, out to higher grazing grounds. Life was isolated from the outside world. The village that Stephen, Charles, and Mark had come from was seventy miles from the nearest post office and the nearest telephone. To get to the road that brought them to Nairobi for meetings with their big-city lawyers, the government, and with me, they had walked five days through the forest.

The great fear that drove them to Nairobi was that their forest was about to disappear because their land, like the rest of Kenya's Masai Mara, was about to be given over to tourism. The Mara—as with the Serengeti in neighboring Tanzania—was already prime territory for tourists on safari. And with tourism also already the biggest foreign currency earner for the country, expanding the park area seemed a no-brainer. The existing lodges were busy, lucrative, and overcrowded. The development plan for the Loita forest threatened to inundate their isolated region with roads. Eight tourist lodges were slated to be built by a South African company. The tourists who would come to the lodges were presumed to want to see elephants and lions, not herdsmen and cattle. So, if the roads and development didn't destroy the life of the Loita Maasai, the pressure for them to move out of the way would. "We would be denied our rights to use the forest," the three men complained, "which we have protected from time immemorial, for our cultural practices and for grazing our cattle."

This would not be the first time this had happened to the Maasai. Once a great and feared people with reputations as fierce warriors, the Maasai ranged over an area that stretched from the Indian Ocean to Lake Victoria, some eighty thousand square miles covering the lands north to Mount Kenya and south to Mount Kilimanjaro. Arab slave traders, even as they were decimating the tribes in all the surrounding country, feared and avoided the Maasai; writer Isak Dinesen described them as "chic, daring, and wildly fantastical."[1] But the first squeeze on their territory and their nomadic roaming came in the nineteenth century when the British colonized East Africa and built the Uganda Railway. The British railway builders forced the Maasai to move out of the way of the trains. They were placed on reserves, like Indigenous people in North America, and pushed further south onto the Serengeti Plain in what is now Tanzania, as well as into Ngorongoro Crater, the spectacular hundred-square-mile volcanic bowl, also in Tanzania.

Then, in the 1930s, civilization caught up with them again. With an eye to the huge tourist and visitor potential, the British colonial government in what was then still called Tanganyika initiated moves to turn the Serengeti into a protected wildlife conservation area. This was finally achieved in 1951. The Maasai became migrants not only because of their tradition of following their cattle, but also because they were pushed to keep ahead of the pressures of development, land protection, and tourism. The half million or so left, wandering back and forth across the territories that straddle the border between Kenya and Tanzania. The pressure was on constantly. In the 1970s they, with their cattle, were evicted from the floor of Ngorongoro Crater in Tanzania. In Kenya, growing tourist development in the Masai Mara likewise made their world smaller and less secure—which pushed folks like the three gentlemen who drank Coca-Colas with me to try to find ways to stay on their forest lands.

The long and the short of our story is that the Maasai did ultimately manage for the most part to stay on their lands. And they did so, in no small part, by becoming part of tourism rather than perceived as being in the way of it. Thirty years after my meeting with Stephen Ole Senteu, Mark Ole Karbolo, and Charles Ole Sonkol in Nairobi, if a tourist from Europe or North America or Asia wants to go on holiday to the national parks of Kenya or Tanzania, the Maasai, when you get there, will be part of the show. An easy web search will bring up any number of safari companies and tours featuring both the Loita forest and other Maasai locales in both Kenya and Tanzania.[2]

The promotions for safaris in these lands of the Maasai are attractive advertisements. On offer are "epic landscapes," "totally unique experiences," "genuine un-contrived culture." Employed is the kind of over-the-top language meant to trigger romantic fantasies in the minds of would-be tourists hoping to put on a pith helmet and turn themselves into explorer Henry Morton Stanley or the like. But is purveying some kind of frozen colonial time anywhere close to reality? Apparently it is and it isn't. Writing in *Condé Nast Traveller*, veteran traveler Steve King opined how modern-day East Africa has come to look a great deal different than when he first visited, given recent decades of population growth, urbanization, road building, and general modernization and westernization. Yet he expressed relief that one could still find corners of "the *old* Africa" if one got away from the cities and especially into the Loita Hills:

> Much of Kenya has changed beyond recognition since my first visit. Parts of Nairobi seem to change beyond recognition from one visit to the next. Here, though, hardly anything has changed at all, or not to the naked eye. Almost 50 years ago, in *The Tree Where Man Was Born*, Peter Matthiessen wrote of the Loita Hills as "roadless and little known . . . that epic Africa of hope and innocence." They are as roadless and little-known today as they were then, except to the Maasai who live among them.[3]

What Steve King alerts us to—perhaps unwittingly—is how the Maasai, instead of being a nuisance and in the way, now have an elemental role in keeping alive an East Africa that is partly in our memories as nostalgia and partly in our imaginations as fantasy. This, fortunate for them, is apparently the East Africa tourists—in their dreams—want to see, and pay up to $1,000 a day to enjoy. The tour companies aim to comply, and if they can't come up with it naturally, they will create just what the tourists want. Maasai Trails safari company promotes the secret of their tours as being

> all about your guides, local Maasai from the Loita area, who are eager to share their customs and culture, whilst teaching you the lore of the bush. The Loita Maasai who lead you on these trails are reputed to be the most conservative of this tribe, dedicated to their traditions, ceremonies and rites of passage. As a people, the Maasai have struggled over the years to maintain their culture and have shied away from large scale developments and lodges that comprise other parts of Eastern Africa . . . thus is the beauty of Maasai Trails. This small low impact camp and mobile camp are a sustainable method to bring income to the local people whilst protecting the natural surroundings of the Loita Hills and sacred Forest, as well as their culture![4]

The epitome of a seductively descriptive ad trading on the expectations of tourists and the performance requirements of their hosts is this from Safari Dynamics:

Day 1–2 Arusha—Maasai Village "Olpopongi"

In the morning we will pick you up from your hotel in Arusha. Afterwards you will undertake an approximate two and a half hours scenic drive along the slopes of West Kilimanjaro up to the Maasailand near Tinga-Tinga. After arriving at "Olpopongi" Maasai Village a group of pleasant Maasai and the Village Chief welcomes you with a chilled welcome drink. During a walk through the village, your personal Maasai guide is explaining the life, past and today's life of the Tanzania Maasai. Get more information about the culinary culture of the Maasai while tasting a traditional tea (chai) prepared on the wood fire. A tasty lunch with a variety of local food highlights will end the activities inside the village. The afternoon starts with an exciting & very educational walking safari across the nearby bush. Maasai warriors will guide you along their paths and explain you typical medicinal plants, extraordinary trees, tracks and the local wildlife. In addition to that, the usage of local tools, hunting techniques and fire preparation is a part your unforgettable bush excursion. Coming back to the village, the well-deserved break is accompanied by freshly brewed coffee & cake. During the sunset a group of Maasai men and woman gathering around for the traditional singing & dancing performances. A great BBQ-Buffet right at camp fire and the "get together" with your already known Maasai friends, will provide you an unforgettable evening under the (mostly) starry sky. The evening will find the end at your private Maasai house on traditional beds with comfortable mats & blankets.

After hearing the rhythm of the bush drums next morning, the breakfast with freshly baked rolls, jam, fruits and an omelette of your choice is waiting for you! Later, during the still early morning, the tour is over and it's time to say good bye with many new impressions in your mind. Your driver will bring you back to your hotel in Arusha.[5]

But there is, of course, a hitch, the hitch being that fantasy is always fantasy. An organization with connections inside the African tourist industry, Tourism Teacher, warns us about being alert to the reality of what we as tourists may be both looking for and experiencing. They use the Maasai as their example, explaining that while a common undertaking for tourists visiting Kenya and Tanzania is a side trip to see Maasai tribespeople, one might be cautiously alert to what is put on display. You can visit a homestead, meet tribal members, and be told about their way of life. But the critics are careful to warn how much of what one might be shown in relation to the Maasai tribe isn't strictly true. What one sees on a tour might well not be how a lot of Maasai people actually live. It is about an idealized past and not the present. Some traditions, the educative website makes clear, are still followed: the Maasai are still a deeply patriarchal society, and the Maa language is still widely spoken. But some of the villages tourists are taken to turn out to be not real villages, but purpose-built stages to provide a tourist spectacle. What is suggested is that large parts of what the tourist will think of as reality is not reality at all, but a show—a show the local native Maasai are obliged to participate in in order to ensure their own very survival.

For the Maasai, the key to participating in tourism has turned out not to just get a modern-day hospitality job, whether that be cleaning rooms or managing bookings, but to be prepared to playact an idealized version of both what their ancestors might have been or, more pertinently, what visiting tourists have in their mind that an idealized Maasai should look or behave like. It is all a form of theater, an idealization in some ways not that much different from our wanting to experience Venice as it might have looked in the sixteenth century, or an English cathedral as it might have been when inhabited by monks. But there is an important question around whether it serves either the purveyor or the consumer well. Is it an indignity for a tribal people to wind themselves back in time—especially if their audience doesn't "get it"? For the audiences, is there a bit of a con game going on, everything charming but kind of phony?

Alexis Bunten and Nelson Graburn explain in their book, *Indigenous Tourism Movements*, that tourists come to Tanzania (or Kenya) hoping to witness a young, male warrior alone on the Serengeti or a woman, bare breasts covered with colorful beads, sitting cross-legged next to a hut. The Maasai comply in part out of economic necessity, but also because many of their traditional cultural activities are now otherwise against the law. Without access to their prior way of life, many Maasai want to be part of the modern cash economy, but they can only do so by projecting themselves as frozen in the past, as part of the touristic *mis en scene*. In a cosmopolitan twist of fate fuelled by the Indigenous imaginary, Maasai hosts are transformed from subalterns—neither allowed to live outside the cash economy nor fully able to participate in it—to ambassadors for the global Indigenous movement's resistance to western consumerism.[6]

In his extensive study in the same book, sociocultural anthropologist Noel B. Salazar, who specializes in tourism and mobility, states in his essay, "The Maasai as Paradoxical Icons of Tourism (Im)mobility," while

> like everybody else, they as a people are well on now into the twenty-first century ways of living and doing things, in the schema of tourism, it has been important that the Maasai remain *Maasai*. That is, that they dress and look and act as visiting tourists expect them to dress and look and act. They must fulfill the fanciful expectations of their visitors and disappoint those expectations at their peril. Overseas tour operators and travel agents (and not only Western ones) keep marketing the Maasai as one of those extraordinary, mysterious Indigenous African communities that have remained untouched by the global forces of modernization. Like a herd of elephants, or a snoozing lion, or the silent Mount Kilimanjaro, the Maasai need to carry the burden of the eternal in order to fulfill the expectations of—and not disappoint—the visitors.[7]

They must be part of the landscape, says Salazar, "like the wildebeest and the zebras." The contemporary reality of the Maasai people—the broader paolitical, economic, and sociological context within which they live and struggle (such as having many of their ceremonies actually declared illegal by their national governments, or being on the bottom rung among the East African tribal hierarchies, often discriminated against in jobs, educational opportunities, political powers, and so on)—is of no interest to the vast majority of visitors. For them the Maasai—and performing as the Maasai of colonialist fantasy—are simply part of the safari package they have paid for.

CHAPTER 9

Indigenous Tourism

*Can Indigenous Communities Get in on the Action and Still
Preserve Their Cultures and Lands?*

THE STONEY NAKODA PEOPLE, WHO PRESENTLY RESIDE ON A NUMBER OF
First Nations reserves in the foothills of the Rocky Mountains just west
of the Canadian city of Calgary, describe themselves as "the original people of the mountains" and explain, on their tribal website, how they "have
continuously used, occupied and possessed our traditional lands since
well before contact with the Europeans. Our traditional territory ranged
from the Great Plains where we hunted buffalo to the Rocky Mountain
foothills and watersheds where we harvested, fished and hunted big game
and over the mountain passes to the British Columbia interior."[1]

Their story, though, has an interesting wrinkle in that a century and
a half ago they were forced to vacate a significant portion of the territory they had traditionally occupied. In 1885 (the same year that the
transcontinental railway, the Canadian Pacific, or CPR, was completed),
Canada's first national park, Banff, or as it was initially called, Rocky
Mountains Park, was created on 6,600 square kilometers (2,300 square
miles) of some of the most beautiful foothills and mountain peaks to be
found anywhere. In preparation—just as had been the threat against the
Maasai in Kenya and Tanzania—the Stoney Nakoda were asked to leave
the designated territory. Perhaps "asked" is putting it too politely. They
packed up their camps and reestablished outside the edges of the park
boundaries.

Fate, however, in the form of a derailment of a passenger train on the relatively new railway, intervened. Lore has it that in the summer of 1894, a CPR train filled with cross-country passengers had to stop around the edge of the park boundary because rains had washed out the tracks. This left the urbane passengers who had traveled all the way from the east marooned in the middle of what many of them considered "nowhere." The train conductor, grasping an opportunity, persuaded the Stoneys, who happened to be camped nearby, to open their tepees and entertain the tourists with food and ceremonies for what turned out to be the next seven days. The initiative proved a great success and thus was born Banff Indian Days which, for the next eighty years until it was halted in 1978, made the Stoneys the cornerstone of tourism in Banff National Park. For a period of a week or so each summer they were invited temporarily back into the lands they had once occupied on the premise that they would set up camp and dress and act as they had in pre-colonial times for the benefit of the tourists. They would pose for pictures, answer questions, and, in general, be colorful.

They were a big hit. Every summer, thousands of tourists flocked to Banff Indian Days, which park organizers billed as an opportunity to see aboriginal people in their natural surroundings. Plenteous photos were taken. The Stoneys enjoyed playing along; in good-natured ceremonies, the chiefs made visiting dignitaries like the Prince of Wales, who showed up in 1919, honorary members of the tribe. Everything, though, was strictly controlled. The tribal people could come to Banff only for the prescribed days that the festival lasted, then they had to leave. They had to behave in conformity with stereotyped expectations as to the kinds of costumes they would wear and the kinds of tepees they would erect. Only certain of their traditional ceremonies were allowed to be performed; for instance, they were not allowed to introduce rodeo even though that had become part of their way of life, nor could they do anything spiritual or medicinal that had been prohibited by the Christian missionaries.

In 1938, things got onto a bit of a collision course when the government tried to limit the number of participants to 150 individuals. It backed down, however, when the Nakodas protested the limitations. According to academic Jonathan Clapperton, who studied the Indian

Days phenomenon, "organizers moved to limit or stop anything that they perceived as un-authentic and participants would have to sign a contract stating they wouldn't drink alcohol, that they'd leave their cars in the parking lot and that twice a day they'd allow visitors to enter their tepees."[2] Banff Indian Days was constructed to temporarily welcome Stoney Nakoda back to the park in a ritual considered safe by onlookers. But that was it. "Indians could not be allowed to run at will through the park as they would ruin it and disturb the balance of nature and destroy all the animals. And at the same time, they were desired . . . by bringing in tourists and boosting revenue to the park."[3]

Banff Indian Days was an early iteration in North America of what we now commonly call "Indigenous Tourism." Its occurrence underscored the sizable appetite among the public and visitors for "safe" insights into the exotica of Indigenous life and culture. However, it left much to be desired in terms of questions like "control" and "benefit," items that are raised justifiably by those currently promoting Indigenous Tourism across North America and, indeed, the world in this third decade of the twenty-first century and certainly as questions raised by the Maasai of Kenya and Tanzania in chapter 8. Recognized as a tourism opportunity by the federal and provincial governments, the Stoney people were given merely temporary passes back into the national park. They could do this but once a year as a spectacle for tourists, just as they were allowed to practice some cultural traditions but only under strict guidelines. The Stoneys were expected to give of their time, open their tepees, and pose with tourists while the economic benefit went hardly at all to them but predominantly to Banff businesses that gleaned the patronage of the hundreds or thousands of visitors.

Indigenous Tourism fits into a fascinating cultural niche in our societies, one that can't avoid being wrapped up inside the myriad folds of colonialism. In travels through Ontario, the province of my birth, as far back as I can remember as a child, my brother and I would watch through the car window for signs of "Indians." When you got away from the urban areas, especially traveling north, you would happen upon reserves, little communities with restrictive-looking signs posted, and at other points along the highways, restaurants, gas stations, and gift shops with suggestive aboriginal motifs like peace pipes or headdresses on their banners. I remember, at one of these, the purchase of a small tom-tom drum, rubber stretched by rawhide laces across a barrel that looked like birch bark. It appeared authentic, but it was not. The barrel was not birch bark but plastic. The tom-tom drum was not locally made. Upon the basis of later research, my tom-tom bore no resemblance to the drums used by the Wendat or the Anishinaabe, who were the local peoples of the area we were traveling through. But the tiny drum bore the image of a chief in full headdress, so to me seemed legitimate. In fact, because of its perceived legitimacy, it is what I, for a long time after, used to judge the authenticity of any "native" artifact. Surely I already knew what was the "real deal."

It is doubtful that the "Indians" living on the reserve next to the highway-side gift shop got any benefit whatsoever from my purchase of this knickknack. I can't remember, but it was possible not one of them was even employed in the shop. This was the case all across North America, whether you were in Navajo country in the US Southwest or in any of the communities along British Columbia's coast. The "Indian," or Indigenous or Native, was publicly presented largely in ways that were stereotyped or even cartoonish. Later, when I started traveling frequently through the Lake of the Woods country straddling Minnesota and Ontario's northwest, almost every little town had a statue of some tomahawk-wielding chief advertising a local resort and every grocery store was called a "trading post." Yet the adjacent communities of "real" Indigenous people were teetering on marginality. Their post-colonial experience was of marginalization and discrimination. Their reservations were isolated and isolating; the rules about their identity were restrictive. Far too often, their economic or social progress into the wider world was dependent not on celebrating their Indigenous identity, but on denying or hiding it.

INDIGENOUS

A generalized definition would term "Indigenous" as people who have occupied their unique territories since time immemorial. More commonly, the term is used for people who have persisted to live—the world over—within lands that have been colonized by outsider settlers. So, the Navajo in America, or the Cree in Canada, or the Mayans in Mexico are well known as "Indigenous peoples." The designation, explain Alexis Bunten and Nelson Graburn in *Indigenous Tourism Movements*, which they edited, "is easier in most of the settler-dominant ex-colonies where racial separations are distinct."[4] The United Nations, through the 2007 Declaration of the Rights of Indigenous Peoples, defined Indigenous peoples as "those which, having a historical continuity with pre-invasion and pre-colonial societies that developed on their territories, consider themselves distinct from other sectors of the societies now prevailing in those territories, or parts of them. They form, at present, non-dominant sectors of society and are determined to preserve, develop and transmit to future generations their ancestral territories and their ethnic identity, as the basis of their continued existence as peoples, in accordance with their own cultural patterns, social institutions and legal system."[5]

Indigenous people the world over have perpetually been subjects of the curiosity of outsiders, especially well-off Europeans who could travel to their lands: Paul Gauguin in Tahiti, Alexander von Humboldt in South America, Margaret Mead in Samoa, Franz Boas on the west coast of North America. The relationships that were engendered rarely embodied fairness or reciprocity, with the worst being the atrocious activities of the slave traders in Africa, the Spanish gold hunters in Mexico and Peru, and the missionaries trying to Christianize North America. The question always lurked whether imbalances could some day be righted, and how this might happen. Could the exotica gap be treated in a healthy way, respected rather than exploited? And the question came up, interestingly, what, if anything, might be tourism's role in this? Could the world's fastest-growing industry perform a positive role? Or was it destined to be just one more function of disequilibrium?

As the tourism industry exploded in recent decades, it seemed natural that Indigenous societies should not want to be left out. There were opportunities that seemed logical for them to explore and exploit. Tourism, for Indigenous societies, write Bunten and Graburn, provides a number of opportunities. Some of them are economic—the chance for a community to earn money and provide employment for its population without industrializing. Others are cultural—the possibility to present themselves and their traditions and arts to outsiders in ways that build bridges and are educational.

The pitfalls are there as well, mainly in the challenges of keeping control of the economic business as well as presenting cultures in ways that preserve the integrity of a people's identity. It is always a fine line to walk. "Indigenous-owned tourism tends to produce popular forms of cultural display, while simultaneously deconstructing it," Bunten and Graburn noted.[6] "Indigenous hosts navigate complex, entangled tourist imaginaries, including their own hopes of what they will gain from the workplace."[7] An example, again, is from our previous chapter: "Tourists come to Tanzania (or Kenya) hoping to witness a young, male warrior alone on the Serengeti or a woman, bare breasts covered with colorful beads, sitting cross-legged next to a hut . . . "[8] The Maasai often feel themselves to be in the same boat as other Indigenous groups engaged in and encouraging tourism: they exist between the rock and the hard place of putting on a show out of economic necessity or resisting, on principal, and missing the economic boat.

The two-edged sword of a tourism economy in Indigenous communities was explored in some depth by anthropologist M. Estellie Smith in a 2010 article for an organization called Cultural Survival, which bills itself as devoted to promoting Indigenous rights and cultures worldwide. After acknowledging the appeal of tourism's possibilities to bring prosperity and jobs to a community and enhance local welfare without the community having to pursue the more environmentally destructive routes of "hard industries" such as mining, forestry, or heavy manufacturing, Smith went on to dig into the potential problems. Many Indigenous people inhabit rural areas which, for one thing, she observed, are generally incapable of accommodating higher-density populations without having to undergo dramatic alterations. In other words, a place can rarely stay the same once tourist visitors and the support infrastructures move in. Small, Indigenous populations are easily swamped, she wrote, when several busloads of tourists arrive or when a cruise vessel pulls into the harbor. Another stressor comes through the expectations visiting tourists are likely to have about entitlement to amenities not likely naturally available in the more-modest existing community. Do they want a swimming pool and a fancy restaurant in their hotel? Do they expect paved roads to get around on where before the roads were dirt or gravel?

Smith then observes what everybody is more and more trying to be alert to, which is the powerful impulse to commodify traditional cultures in order to make them saleable to the tourist visitors:

> In order to maintain or increase the flow of tourist monies, a contrived and artificial "folk culture" is frequently staged. It may even be invented as locals respond to the expectations of visitors. Townspeople will invent a "traditional blessing of the fleet" to attract visitors; and "old town" will be newly built (complete with hired locals dressed in the costumes of some archaic and often mythical "olden times"); new "folk crafts" will be designed and taught to the natives by outsiders.[9]

HAWAII

The issue of local populations being left in the dust (or crushed) when the tourism industry steamrollers through seems even more acute when those locals are Indigenous peoples, living on lands long occupied by their ancestors but then overwhelmed by visitors. In an article about Hawaii, writer Cristell Bacilio reports how deep run the feelings among native Hawaiians about their fate in a territory where tourism is a huge driver. After statehood in 1959, tourism exploded with Hawaii becoming the place to go, especially for a winter holiday from the mainland. By 2020, ten million tourists a year were outnumbering the Hawaiian local population sevenfold. For a good number of these locals, the story of tourism is that it had got away on them. This is especially the case for native Hawaiians who, numbering 310,000, make up about 25 percent of the state's permanent population. "Tourism's takeover of Hawaii," wrote Bacilio, "displaces the native population, intrudes on their sacred land, and interferes with each household's economic stability, regardless of whether the person works in the tourism industry or not."[10]

Bacilio cites a 2022 survey carried out by a Hawaiian television station, KHON2, in which 67 percent of residents agreed that Hawaii is being run for tourists at the expense of local people. The survey listed the major problems traceable to tourism as including overcrowding, damage to the environment, and the higher cost of living. "Tourism has caused environmental damage to the land and water sources, which has led to the ongoing water crisis. The residents not only have to deal with the limited water supply but also the high cost of living."[11]

UPDATE ON BANFF

In Alberta, Canada, after a lag of more than thirty years, the Stoney Nakoda were back in Banff National Park. The return, in 2004, came at the initiative of a then fifty-four-year-old elder and artist, Roland Roll-inmud, who worked with Parks Canada officials to engineer something that would be enticing to tourists and, at the same time, legitimate in the eyes of the Stoney Nakoda people. Banff Indian Days henceforth would not look like the old Banff Indian Days, but would focus on Nakoda culture. It hoped to be seen as an opportunity to help Nakoda youth learn more about their heritage.

By 2023, a great deal had changed in Banff: There had come to be a host of Indigenous-themed and Indigenous-controlled events available throughout the year in the park and in the various culture and arts centers in the Banff town. These were accessed by tourists and Nakoda people alike, especially during the month of June, which had been declared Indigenous People's Month.[12]

MANITOULIN ISLAND

Across North America and indeed the world, all manner of groups and communities have decided that Indigenous Tourism should henceforth be spelled with a capital *I* and a capital *T*. Employing slogans like "our story through our lens," as does one community I encountered, they are trying to take charge of the economies of tourism along with its content both philosophical and cultural. Welcoming, and indeed planning for, tourists to come visit them, a rapidly growing number of Indigenous folks are trying to transform themselves and their communities from passive objects into active subjects.

One such initiative can be found on the Canadian island of Mani-toulin, or, in the Anishinabek language, Odawa Mnis. *Manitoulin* is an

Algonquin word translated roughly as "Spirit Island." At 129 kilometers (80 miles) long and up to 48 kilometers (30 miles) wide, it is the world's largest island in a freshwater lake, the lake being Huron, the third largest of the five Great Lakes. Manitoulin sits at its north end, facing toward the US state of Michigan and separated from the Ontario mainland by the lake's North Channel.

Created by both geological and glacial actions, Manitoulin is walled on one side by the escarpment, the grand fault that runs all the way south to Niagara, and is also part of the rim of something called the Michigan Basin, a geological saucer underlying southern Ontario and much of Michigan.

From the standpoint of the visitor, the island has a pleasantly unresolved feel, not quite pastoral, though it might be, and not quite rugged shield, though it comes close in places. It is fitting perhaps that the entire east end, about 160 square miles, or 412 square kilometers, is Wikwemikong (aka Wiikwemkoong), the fifth-largest First Nations reserve in Canada and, in the absence of ever having signed a treaty with any government, persistent in hailing itself as "un-ceded territory." In total there are six First Nations, with 6,500 residents, or about half the island's 12,800 year-round population. The smallest, Sheguiandah, has only 130 people, while more than 3,000 live at Wikwemikong.

The Ojibwe, aka Anishinaabe, are the predominant Indigenous peoples who have inhabited the island for at least a thousand years. The arrival of Christian missionaries (Jesuits and Recollects sent by Quebec governor Samuel de Champlain) and European settlers in the 1600s brought disruption and disease which, along with wars between the resident Ojibwe and the Iroquois, who raided from the south, caused substantial numbers to flee temporarily as sort of refugees, some going as far as Quebec. A century on, more European settlers arrived to clear the land of timber and start farms and fisheries which caused the Indigenous who remained, as well as those who had returned, to endure the usual pressures from land grabbers and governments wanting to gather them onto reserves.

The notion of "spirit" is strong in both the identity and the cultural practices of the Anishinaabe who inhabit the island today. Their

CHAPTER 9

foundation myth stories consider Manitoulin to be the special place ordained by Kitche Manitou as the best piece of creation and the place where Nanabush, a spirit in nature as well as a trickster, was given freedom to operate. Even the non-Indigenous visitor can get into the spirit of this (no pun intended) by paying attention to a kind of shimmery mood that hovers over the island, especially when you walk along some of the shorelines or climb the inland hiking trails.

The non-Indigenous, or settler towns and villages of the island, support themselves with a bit of lumbering, some fishing, and a lot of farming that ranges between spreads that look blissfully prosperous to others that appear to barely subsist. A kind of steady though not ostentatious tourism has always supplemented these activities to bolster the economy. The summer population of the island expands by 25 percent, or about three thousand extra people, mainly summer cottagers and boaters who all spend some holiday money. A couple of well-appointed marinas on the North Channel at Gore Bay and Little Current stay busy with both American and Canadian sailors and yachtsmen, and a number of trailer parks and cabin rental facilities are also occupied near to the island's beaches. A variety of attractive and well mapped and marked hiking trails take walkers past waterfalls and cliffs, all part of a designated route called the Great Spirit Circle Trail. Fishing in the spring and summer and deer and waterfowl hunting in the fall are attractions that bring in quite a number of outsider sportspeople.

It's not that there's been no tourism on Manitoulin supported by First Nations folk. Neil Debassige, a member of the M'Chigeeng First Nation, where he was also principal of the school, operates a cottage business started by his father forty years ago. Neil, his wife Dianne, and two daughters have expanded that to a family-run business of cottage rentals and salmon, bass, and trout fishing excursions. The cottages are grouped in the middle of the island, on the shore of Lake Mindemoya. The fishing expeditions run out of two marinas, one on the north side of the island and one on the south. Neil will take you out in his twenty-three-foot Sea Swirl boat, make sure you catch some fish, and then fry up a tasty shore lunch. The Debassiges entertain about a hundred guests a year in a season that runs from June through September. Fishing guests have come from as far away as Germany and the Netherlands, though most are from nearby southern Canada and the northern United States. But he laughs when I ask whether he considers his to be capital *I* and capital *T*. Indigenous Tourism with a capital *I* and a capital *T* comes up not so much from individual operators like the Debassige family, but on a more official level from First Nations governments.

On Manitoulin, the more ambitious and, one might say, pioneering of such enterprises are those being put forward by Wikwemikong. Wikwemikong Tourism is the source of the slogan I previously mentioned: "our story through our lens." On the phone, I am told by Olivia that if I want insight into the tourism initiatives of her community, I should show up for an open house due to be put on by a number of local services. Olivia carries the title of tourism coordinator for her community.

So, on a July morning, my wife and I leave the cottage we have been renting for a week and drive across an hour's worth of Manitoulin. We proceed over winding roads through forests and past farms to turn off at last into the Wikwemikong Unceded Territory (announced by a substantial sign). Continuing our journey, we crest hills to be rewarded with splendid views of stretches of placid water and inviting shorelines. Eventually the evidences of settlement appear: a house set back from the road here, an auto repair shop there, and a Roman Catholic church, St. Ignatius. Then, buildings come at us more frequently and look more official: a health center, an early childhood center, a seniors' residence, a building for the tribal police.

I have been told that the open house is to be set up on the baseball diamond. "Just go where you see a bunch of cars," a young woman on a motorbike tells us when we pull up to an octagonal red and white sign that says NGAABZAN instead of STOP. We locate the ball diamond with its plethora of cars and a vast circle of tables under awnings. Signs advise that the people behind the tables will be able to provide information on just about every project and undertaking the community might want to have: early child care, policing. A fellow named Theodore, promoting the welfare of birds and a scheme to repopulate the area with bats, takes us aside to ply us with his brochures. At last, sure enough, there is a banner for Wikwemikong Tourism and a table piled high with more brochures. Behind it stand Olivia and a couple of her associates.

Wikwemikong, known colloquially as Wiky, has already been hosting a major touristic event for more than sixty years. That is the community powwow, now called the Annual Cultural Festival, that happens over three days each year in early August. Theirs got its start in 1961. In Indigenous communities across North America, powwows are grand cultural festivals that gather people from far and wide for dancing, feasting, craft exchanges, and ceremonies. For theirs, one of the largest in North America, Wiky will see many hundreds of people, mostly other Indigenous people coming from far and wide.

For the purposes of targeting non-Native guests, "tourists" in the more conventional sense, however, Wiky is trying other ventures. Their aspirations, as put forward on their website and laid out in a rather handsome ten-page booklet, revolve around a combination of local nature—bits of which are quite spectacular—and presentation of their historic culture. By historic, they mean basically pre-colonial, some semblance or recollection of what things were like before the arrival of the first white settlers or missionaries. For example, you can sign up for a "pre-colonial culinary experience." Individuals, families, and groups can pay $200 per adult or $515 for a family and then spend the day with a guide foraging in the woods and the waters for what will turn out to be your supper. This could be "wild edibles" combined into a venison dish "cooked on a flat rock over the open fire." Or, natural plants seasoning fresh-caught trout cooked over a cedar plank drizzled with fresh, locally made maple syrup. "Complete your dish with wild rice, and hot or cold herbal teas." After dinner, guests might learn how they can clean their teeth using red willow twigs with spruce gum employed as a toothpaste. The spruce gum has both "antiseptic and disinfectant properties."[13] Other outings on offer emphasize the landscape and scenery. Guided camping trips can be arranged into eighteen thousand acres of territory the community owns across on the mainland, a place called Point Girondine Park. Fishing, boating, and canoeing outings can also be booked. A staff of locals perform as guides for all these ventures.

There are a number of ways one can get to Manitoulin from the rest of the world. By car from the north over a picturesque swing bridge at Little Current; by car ferry from the south, which means a two-hour ride on the *Chi-Cheemaun* from the mainland town of Tobermory. Or via one of the numerous lake cruise boats that dock in a couple of the island's harbors from where, for those who have made previous arrangements, Wiky will send its minibus to pick you up. Hope Opawamick, a voluble and friendly summer student who is in her fourth year doing the job, tells me that she likes the tourists and likes working with them. "Tourists have empathy; they really want to learn, to see what we have seen through the millennia." For the current season, things are just getting started; in fact, I encounter Hope when she is only on her second day on the job. But previous years have been interesting and busy, she tells me; they have had visitors from as far away as Peru.

On another day, we travel in the other direction and stop in the parking lot of the Ojibwe Cultural Foundation on the M'Chigeeng First Nation. The foundation's home is a striking eleven-thousand-square-foot octagonal building with a soaring tepee-like roof, inside of which is a combination of gathering and performance spaces, art gallery, museum, offices, and workshops. You can find it just up the road from the community center, school, gas station, grocery store, and well-kept Virgin Mary grotto outside the Catholic Church of the Immaculate Conception. Created in 1974, the foundation is the product of six First Nations across the island and has as its board of directors all their chiefs. A look at the guestbook tells us we are the twelve and thirteenth visitors to have come through the doors on this day, preceded, an hour earlier, by a couple from Austria.

While a friendly elder in charge of workshops shows my wife jewelry incorporating porcupine quills, I sit down with Naomi Recollet, manager since 2019 of collections and archives. Naomi explains that despite visitors like us, the center is not really intended as a tourist destination—it has just turned out that way. It was set up for the purpose of protecting local culture through such activities as salvaging objects, protecting local knowledge, and providing language classes. Elders serve as an ongoing advisory circle. Yet outsider visitors amount to a couple thousand a year, according to Shaelynn, also last name Recollet, a curatorial intern. Many are drop-ins, though occasionally there will be a planned visit such as a recent group of sixty sixth-grade students who came by bus all the way from Toronto. Most visitors spend an hour or two and treat the experience as one would a visit to an art gallery or a museum, though some will sit down with Darlene Bebonang, as did my wife, to work at fashioning some small souvenir (animal products used in these enterprises all come from donations of carcasses that are found objects, mainly roadkill; the poor porker that was source of the quills for my wife had sadly met an automobile in an untimely fashion).

There is no entry fee to the center because, as Naomi says, they have never conceived of themselves as a money-making venture. Hence they have never figured out how to turn tourists into dollars—though interest from guests is certainly there, and people used to paying museum entry fees wouldn't bat an eye to do so here. So far, however, that is seen as contradicting a basic dedication to cultural authenticity and reclaiming what was lost—or was at high risk of being lost—through colonization. "Our goal," Naomi stresses again, "is protection and presentation of our art and culture, firstly to our own people and our own children, and to provide a safekeeping place for art and artifacts."

When I ask how they handle preconceptions they might encounter from their tourist guests, Naomi says they consider it part of educating. She refers to visitors who arrive without much knowledge of Indigenous peoples as "fresh off the boat," which, in the case of Manitoulin, means they might have just an hour earlier arrived on the two-hour car ferry from the mainland in the south. "People sometimes expect buckskins or ask, 'Where are the tepees?' Or they ask, 'Where are the real Indians?'" At which she simply points to herself. "We try to educate by stressing that those are old ways, but it's now the twenty-first century." They continue to wrestle with how to handle outsiders who are curious and sincerely so while providing authentic information and putting the past into context with the present. The philosophy of the elders, she tells me, is to be cautious about what and how much cultural lore—including about medicine plants or ceremonial rites and so on—gets shared past their own community members.

This issue is referred to as "cultural integrity," and cultural integrity is certainly on the mind of the CEO of the province-wide organization Indigenous Tourism Ontario (ITO), forty-six-year-old Kevin Eshkawkogan. Eshkawkogan, who just happens to be from Manitoulin, a member of the M'Chigeeng First Nation, recalls an early motivation to get involved when, twenty-five years ago, he watched tour buses arrive in his community. "Although we'd been guiding visitors on these lands longer than anyone, we were excluded as tours would come through without First Nations guides. The outsider presenters would tell guests their versions of Anishinabek culture—lots of it not true—and refer to us as 'our Indians.'"[14]

The ITO was brought together in 2008 and echoed what was/is happening across North America. Purveyors of Indigenous Tourism everywhere are striving to think through and organize, as ITO's mission statement puts it, to "assist Indigenous communities and entrepreneurs with programs that build capacity to produce high quality products and services."[15] The organization has a small staff and offices on the Aundek Omni Kaning First Nation and in Toronto.

Indigenous people think differently, Kevin explains. "We operate more communally and cooperatively than does the rest of society with its capitalist entrepreneurial models." The infrastructure they conceived reflects this. One of the spots where you can stay on Manitoulin is a fine modern conference hotel that looks out over the waters of the North Channel at Little Current. It was developed and is jointly owned (as with the Cultural Foundation Centre) by all the First Nations on Manitoulin. "I tell schoolkids, this place is yours," Kevin says, "and their eyes light up. They can't believe it."

The tourism organization's guidelines around cultural authenticity, supported by advice from a specially created Indigenous Cultural Advisory Group, identify things that need to be taken care of. According to Eshkawkogan, when Indigenous communities offer cultural experiences, they should include elements that showcase the culture and educate visitors. "Sometimes a tour bus loaded with visitors will approach wanting only a dance performance," but he urges his members to resist such compartmentalizing. They should be pressed to be more holistic.

Communities also need to stand firm in differentiating between the spirituality of their culture and the putting on of a "show." "A community can't sell a sweat-lodge experience to an outsider tourist," he says, "but they can educate visitors about sweat-lodge ceremonies. They can't give out medicines, but can tell visitors about them." He is optimistic about the intentions of his guests. "Individual visitors are frequently blank slates," he says, "open to learning about our communities in real ways."

The ITO mission is to oversee cultural integrity, marketing and branding, and work to simultaneously develop a skilled workforce as well as an infrastructure within which those persons who have been trained can find real jobs. These are potent matters for communities like Wikwemikong and M'Chigeeng. Being tourist-ready, physically, and having a philosophy of what you want to present are significant issues. The Ojibwe Cultural Foundation Centre, by not having defined itself as a tourist site, is caught having to rethink or at least respond to the idea that it might be just that. Will it charge an entry fee? Wikwemikong has a brochure and a website and thought-through programs, but seems in a scramble to have the trained staff to deliver those along with other amenities such as enough accommodation should the numbers respond to their invitations.

The ITO assesses that in Ontario, a province with a population of fourteen million, Indigenous Tourism contributes over $665 million CAD to the GDP. That is the amount of cash that comes annually out of the pockets of tourists for what they define as "Indigenous" products or experiences. Eshkawkogan calls Indigenous Tourism the fastest-growing tourism sector in Canada and says that the main audience has shifted from mostly foreign Europeans to domestic visitors: "One in three domestic tourists now profess an interest in an Indigenous experience." This seems to be duplicated across North America, meaning that Indigenous Tourism is certainly on the map.

CHAPTER 10

What Digital Photography Has Done to Tourism

IN THE SPRING OF 2007, I WAS IN DUBROVNIK, THE FABLED CITY ON THE Croatian coast of the Adriatic. I had spent an hour, under a very warm sun, tramping the two-kilometer distance of walkway that stretches along the top of the city walls. Built in the thirteenth century, the walls, with their paved stone walkway, are an irresistible magnet for visitors like me. Marching along, I could gaze out, on the one side, over the dazzling blue of the sea while on the other, I peered down on the red tile roofs of the old city. The city is truly old, having been founded back in the eleventh century. Once back down at ground level, I threaded my way through narrow streets, zigging and zagging around the square until I found a small table in a half-covered-over cafe. There I ordered something to drink.

I was hot and exhausted and the shaded cafe was cool. The waiter delivered a lemonade. Across from me a group of fellow tourists, three or four as I remember, were having a great deal of fun. I can't recall where I thought they might have been from, Germany or the Netherlands possibly. The point was the fun they were having, which took me a couple minutes to figure out. Finally I realized what they were doing that was producing so much pleasure. They were taking pictures of one another and then, turning the camera around, were showing the results, all amidst gales of laughter.

The digital camera dates back to 1975, reputedly the invention of an Eastman Kodak engineer named Steven Sasson. Relying on CCD (charged-couple device chips), Sasson's creation, albeit a big clumsy thing by current standards, stands as a prototype for everything we now have in our compact phones. It took awhile. The original, produced in the laboratory in Rochester, weighed almost nine pounds and was about the size of a current-day microwave oven. It required sixteen batteries to make it go, and the process of taking a digital picture took twenty-three seconds.

In 2007, I had a digital camera. The thing, though, is I still thought of it as something used to take pictures in pretty much the traditional way. Carefully and with forethought. What would become the pervasive activity of selfie-amusement—the using of the camera as a toy—had escaped my notice until that moment in the Dubrovnik cafe. This device, it turned out, wasn't simply for mundane work—it could provide a lot of amusement just horsing around and sharing the results.

After that, it didn't take long to note how the practice had become a phenomenon. You needed to keep your camera constantly out and handy. Then, a half decade later, another big change occurred when our cameras moved from being discreet machines to their new incarnation as aspects of our mobile phones. At that point, combined with another tourism moment, I was introduced—on a trip to Rome as it happens—to the next-generation refinement: the selfie stick. Down from the Forum and all around the Colosseum, crowds of young people were milling about, not consulting their guidebooks but busily and joyously positioning their iPhones high above their heads and snapping themselves. They were doing this in the milieu of—though possibly oblivious to—all the ancient surroundings.

Of everything that has had an impact on our act of traveling and tourism, the digital camera/smartphone has to rank pretty high on the list. People always took pictures. But the difference between the roll of Ektachrome color-slide film in your SLR and, by contrast, holding up and touching your phone doesn't even bear comparison. With your SLR (and before that your Brownie Instamatic) you carefully considered the picture you were going to take. Film was expensive, and there were only twenty-four or maybe thirty-six shots on the roll. You looked at your subject. You checked the light. You considered the value of your subject in the overall program of your trip record. You probably felt you needed the Eiffel Tower even though there were dozens of images available at any nearby postcard kiosk. But after careful framing, you probably took a shot anyway. But just one. Or maybe two. Then you waited three weeks until you returned home and had the film developed to see if it turned out.

When I was a kid living in a small, rural community, anybody's travels became a shared experience through the vehicle of their pictures. In the church basement, in the schoolhouse, in grandpa's living room after Christmas dinner, Aunt Margaret's bus trip to California, with a stop at the Grand Canyon, became a public event, a show to appear on the big screen courtesy of her slide projector. We had not yet reached the stage of rolling our eyes and looking for the exit when somebody brings out their trip pictures. For those of us who had not or not yet gone anywhere, anybody's travel adventure could become ours, through the vehicle of the pictures. The fact that Aunt Margaret had seen the Grand Canyon through her own eyes and was sharing the evidence with us pictorially was significantly different than looking at a picture of the Grand Canyon in a school geography book. A couple levels of immediacy had been transcended along with, for the younger among us, the stirring of possibilities. If Aunt Margaret could do this, why not me? And maybe even further than the Grand Canyon. But I'd need a good camera.

The most exotic travelogues we got to share in came via some missionary who would appear at our church in order to give us the lowdown on the work they were doing in Korea or Sierra Leone. This type of travelogue as well as this kind of traveler was a harbinger of the do-good tourism to become a big deal later, in our present day when privileged schoolchildren from North America go off to make adobe bricks to erect a health clinic in Peru, or retired folks pack work clothes into the bag because they are going to participate in a water reclamation project in Ethiopia. The returning soldiers—the missionaries in those days—were awe-inspiring indeed. But they'd have been nothing without their pictures. And what the pictures primarily kindled in us was the notion that these faraway places were indeed possible. Whether or not we harbored any purpose about converting the heathen, a lengthy sea voyage or a complicated (and expensive at the time) set of air connections and it could be us standing in the grass-thatched village light years away from the world we had come from and were familiar with. These possibilities all came to life through the power of the photographic image, the seduction of that image having been created by a person who was now not abstract, but standing right in front of us. Possibilities can be a powerful thing.

Are travel and the camera mutually interdependent? It is doubtful that any general tourist in 1830 would have been lugging a clumsy daguerreotype camera along with them. But after the 1901 development of the Brownie by Kodak, that would have changed dramatically. I wrote earlier in this book about the Asian tourists (I'm afraid we like to use Asian tourists almost as a cliché when it comes to picture-taking) jumping off the bus in the middle of Oxford and scrambling to take pictures of themselves and one another. But a phenomenon I've not quite come to terms with is the one that tells us that getting the picture is all that is needed.

I like to watch people in art galleries. I find it fascinating to observe human behavior when we are confronted by something that either moves us or we have been told ought to move us. Thus, gallery-goers in front of great paintings are good for that. Some, visiting solo, take a great deal of time to examine, ponder, and think. Others, traveling in pairs, discuss and compare notes. Of course, there are those, usually young art students, who unfold their portable stools and try to sketch a copy of what they are looking at. What I note more and more, however, is the phenomenon of visitors moving through an exhibition and—now that it is widely permitted to do so—using their iPhone to photograph the art, sometimes not even slowing down while doing so.

My daughter tells me about visiting MOMA in New York and observing that very phenomenon as people passed Vincent van Gogh's *Starry Night*. Now, one would think of that as a picture you might want to stop and spend a few minutes with, not just look at it on your iPhone when you got home. If you even did that. But no, her count tallied an astoundingly high number of drive-bys. It is the same when they/we are out on the street being tourists. Some of us stop and absorb all that we are confronted with; many more snap and move or snap while moving. What's going on?

In her 1973 book, *On Photography*, Susan Sontag wrote: "Photographs are a way of imprisoning reality. . . . One can't possess reality, one can possess images—one can't possess the present but one can possess the past."[1] This observation provides a substantial insight into our act of traveling and having the camera along with us as a tool of necessity. It is a necessary item for your trip every bit as important as your toothbrush or a good pair of walking shoes. Sontag was speaking, of course, about a different photography than that done via the smartphone. Nineteen seventy-three, when she wrote this, was twenty-five years before digital cameras were in common use and thirty-five years before the iPhone became a camera. Sontag, who died in 2004, wouldn't even have experienced the latter, but you can bet she would have had something to say about it.

Traditionally, of course, the camera—and this was the type Sontag was thinking about—was the implement that you used to provide for later enjoyment. Via your camera and its film you brought your trip home with you and relived at least the selected parts of it later, seated comfortably in your own living room. Hence she observed: "Needing to have reality confirmed and experience enhanced by photographs is an aesthetic consumerism to which everyone is now addicted. Industrial societies turn their citizens into image-junkies; it is the most irresistible form of mental pollution."[2]

This, as well, gave people pause in 1973 and should even more so fifty years later. Which brings up the question of *not* bringing a camera or a smartphone with you on your trip. What are the implications of that? Why would you do it? Would you even think about it?

I had a friend (now deceased), Neil, who during his lifetime traveled all over the world. Fifty-one countries, he once boasted. In the closet off his living room Neil maintained the evidence of those trips, dozens of boxes of Kodachrome and Ektachrome slides. I never sat through any picture shows, though I did note that the boxes were carefully labeled, country and year. I was not aware that he spent any time looking at them himself or that they were shown to anybody else—an evening with popcorn and a projector. But they existed. They took up a lot of space, both physical and, perhaps, psychic.

In the taking of photographs are we fulfilling an obligation? Is it a simple statement, a recording of the fact: "I was here"? It's a fascinating and indeed puzzling human behavior, something that on one level is an act of collecting, a fetish many of us indulge around numerous objects or experiences. On another level it is affirming; that is to say, my taking of this picture affirms my existence at this moment and in this place. If it is used to record iconic locales, like the Eiffel Tower or the Roman Colosseum, that are popularly confirmed as such, then the affirmation is even more significant. "I was here, in this generally agreed-upon import-ant space," the subtext being that I am now, little old me, part of history. In the same way as "I passed by" in front of the original of the painting *Starry Night*, the pic will confirm it.

Interestingly, for the last few years of his travels Neil did stop taking a camera with him. He continued to travel—to Uruguay, to Shanghai—but he went alone, and camera-free. It got in the way, he said, of what he really wanted to do on his trips, which was meet the people.

Then one afternoon when I visited, he told me as well that the closet next to his living room was now empty. I looked at him quizzically.

"I threw out my slides," he said.

"You had thousands."

"Yes, it was all far too much."

Abruptly, in his approaching old age, Neil's vast collection of images somehow offended him. Or seemed to have become a burden too great to bear. Did anybody want them? For whom would they have meant anything?

Why is it that we take and then collect photographs? What is so important? Is the moment of shooting the important one? If that is the case, then the digital camera and its offspring, the smartphone camera, is the perfect vehicle for a culture like ours. The fun of the smartphone cam-era is all in the moment, the moment of pressing the button and then, possibly, the moment of transporting the image forward to Facebook or Instagram. In those places it will have an even more transitory existence; at best it will achieve a number of "likes." And that will be that.

As I write this, my wife is in Germany. It is a trip she took partly to visit relatives, partly to see the sights. Stephanie has made it into a couple of museums, a couple of castles. She travels well on her own, though it has become apparent to me that this is a different type of traveling. That is because she is not really doing this on her own. Every few hours, I get a text. With pictures. Her Facebook and Instagram friends get even more, all of it pretty much instantly. She made use of a high-speed train the other day. We were all privileged with the sight of her boarding the train. The upshot was that it brought her trip instantly to us. It wasn't like the old days and seeing the stuff weeks later. Or like the missionaries from Sierra Leone, months later. It was just that morning! And the fact that neither she nor the rest of us is likely to ever view that picture of her boarding the train again is irrelevant. The events happen, but the evidence is pretty much disposable. Or the evidence is so overwhelmingly voluminous that it gets lost in our archives. How many pictures do you think any of you have on your phones? Have any of them been carefully collated?

With our generation's novel technological possibilities, are we watching the transition into something truly original? On a number of levels, I believe we are. Time, for one thing, is being altered. I, sitting back at home, can experience my wife's travels pretty much in real time; I don't have to wait for that later report. There is no way and little need to stretch things out. The second item on the table is sharing. If Stephanie wants to make it so, it will no longer be just her trip, it can be my trip too or the trip of every one of her Facebook friends. One day, we received a picture of her lunch—a wurst and a vegetable soup. Our technologies not only allow us to share, they invite us to share.

Lastly, there is perspective. Instantaneous digital photography has made both us and the world infinitely more narcissistic. If we want to be generous, we might term it self-referential, narcissistic if we are being less kind. Either way, the relationship between ourselves and the objective world has undergone a profound transformation. The focal point has changed. Yes, people always had their pictures taken standing in front of Niagara Falls or the Eiffel Tower. But now the photographer is—infinitely more often than at any time in the past—the subject of the photo. Standing there front and center, with the Great Wall of China or the double-decker London bus or the pyramids of Giza being merely the background. *We* want to be the picture; the commercial postcard industry featuring merely objective images of any site has pretty much fallen by the wayside. Taking photos of oneself has possibly reached to the level of a disease, an epidemic for which there is no vaccine. Our experience of the world has morphed into almost completely subjective rather than objective; the digital technology provides for immediate rather than delayed gratification. And we are hooked.

CHAPTER 11

Tourism Futures

IN SEPTEMBER AND OCTOBER OF 2022, *CONDÉ NAST TRAVELER* MAGA-
zine devoted its entire issue to "The Future of Travel."[1] There have always
been a slew of "travel" magazines, most of them filled with gorgeous
pictures and breathless articles basically promoting rather than critiqu-
ing the activity of global travel. Some fall into niche categories—there
are magazines that are only about eco-tourism, or adventure tourism,
or service tourism. But promotion, rather than examination, is the
main editorial philosophy of virtually all of them. Every airline in the
world has its own in-flight magazine built on that premise. And, for
thirty-five years, Condé Nast has been one of the more glossy and yet
easily read. Do you want to go to Rotterdam? Amsterdam? Goa (India's
"sunshine state")? Would you like advice on where to stay when you
visit Yellowstone National Park—at any season of the year? How about
a tasting trip to sample wine and food in Italy's "discreet, sophisticated
Piedmont region"? For all such publications, and especially for Condé
Nast, a tumult of ads pays the freight. So the act of devoting the entire
late 2022 issue to the future, while not inconsistent, was at the same time
munificent and thoughtful in its premise.[1]

What, according to the provocative thinkers at Condé Nast, are the
elements of travel's future? Well, there was an acknowledgement of the
impact the Metaverse is likely to have. Interestingly, the conclusion drawn
by writer Toby Skinner was that those virtual visits via your headset to
Machu Picchu or Mount Everest will, rather than sate us and dampen
appetites for any "real" travel, in fact whet those appetites. Something

deep inside us will realize that virtual is woefully inadequate. "I can't help thinking," Skinner wrote, "that in so many ways, travel is actually the antithesis of an ever-expanding digital universe; that travel is still the ultimate celebration of the real world—nature, people, connection, human experience in all its richly messy complexity—and that one of its key roles in the future will be as an antidote to cyberspace. . . . For all the metaverse talk of community and connection, my default online mode is a sort of isolated, vague misanthropy. Real people tend to be intriguing, surprising, and kind."[2]

There might, of course, be space travel in our futures, certainly if Richard Branson or Elon Musk have their ways.[3] But here on earth, services like hotels will try to be significantly different than their up-to-now models. For a while, the greening of the hotel industry was simply small-gesture stuff; I recall, in an early ripple, how a room in a Sheraton hotel twenty years ago would include a bold-lettered card advising me to "pitch in to save our planet!" The accompanying instructions stated that were I to sort my towels each morning, it would give housekeeping staff the signal to launder only the ones I need. Doing so would "help save thousands of gallons of water and laundry detergent." Another exclamation point. I realized later that while this possibly helped the planet by saving some water, what it did mainly was save the hotel a substantial amount of labor and expense in their laundry rooms.

Latterly, conservation, sustainability, inclusivity, and community have started to become terms with more teeth. In the hospitality industry, resorts in Mexico might have their own solar farms to produce their energy; hikers in the mountains of Kyrgyzstan will sleep in yurts so as "to leave the terrain practically untouched." Camps in Kenya's Masai Mara will rent (rather than expropriate) land from Maasai tribal peoples, giving families and communities reliable sources of income and negotiating to conserve parcels for livestock grazing and transit corridors for wildlife. Resorts in the Dominican Republic will imbue the stays of their luxury guests with fishing programs, river cleanup, and community education.[4]

On the very day I am writing this, my email inbox included a promotion from a hotel chain whose resort in the Dominican Republic I'd once stayed at. Over the signature of the chain's chief sustainability officer,

I was cheerfully advised that due to their henceforth more responsible management of energy, water, and waste resources, my future stays would be not only comfortable but also respectful of the planet. I was also promised "a healthier environment, quality local products and authenticity in every service." The results of the initiatives will make us all happy. The Great Barrier Reef will be protected, in part because tourists with their cameras (voluntourists) will join a raft of other interested parties, such as yacht sailors and fishermen on their boats, in photographic documentations of the state of the reef. The data gleaned in this annual "census"—thousands of images—will inform the actions of marine scientists in assessments and strategies for damage amelioration and repair.

In a radical makeover of aviation, electrically powered airplanes may well become more and more the norm, at least for short-haul flights, and sustainable aviation fuel (SAF, produced not from petrochemicals but from such sources as forestry and agricultural waste, used cooking oil, carbon captured from the air, and green hydrogen and touted as the key to reducing CO_2 emissions by up to 80 percent), once it overcomes problems of being cost prohibitive and in short supply, will be common for longer hauls. That will be unless, in a rapidly changing landscape, both of these get beaten to the punch by zero-emitting hydrogen fuels. Such innovation, which might come more quickly than we think, would make redundant the prediction of Swiss bank UBS that a quarter of commercial planes will be electric or hybrid by 2035.[5]

Down at water level, cruise ports will make efforts to go green; the CLIA, Cruise Lines International Association, has predicted net-zero emissions for its oceangoing members by 2050. This they will do by enabling the big ships to plug in to onshore power at every port of call. This will require both plug-in facilities in the world's major ports and capacities on the vessels themselves for electrification of their propellant power. Plug-in capable: apparently it's on track to happen. The list goes on. Eco-hotels will be more numerous and easier to book. Aborigines in Australia (and Indigenous peoples everywhere, as with the aforementioned Maasai) will have more ownership, input, and control as to how their lands and cultures are visited.

These are ideals, and appealing ones. Taken together they speak to an increasing thoughtfulness in populations that are now well into second and third generations of environmental thinking and increased cultural sensitivities. However, against the ideals, there are still the powers of economics and sheer numbers. For every good and hopeful thing that might happen, what if ten negative things continue to outweigh it? There is the challenge.

As it has with many other things, the rather solemn notion of "responsibility" has come to infect our language around tourism. Traveling responsibly, like recycling your trash or ceasing to use plastic shopping bags, is a principle we are urged to assent to. Doing so will make us feel a bit righteous, as if we are somehow putting our shoulders to the wheel of that enormous task Western humanity—at least the "with it" segment of it—has set for itself: saving the planet. In the field of tourism, environmentally responsible travel got formalized a couple of years ago by something called the Glasgow Declaration on Climate Action in Tourism that came out of the COP26 (Conference of the Parties) meeting in Glasgow, Scotland, in 2021. That conference, under enormous pressure like so many of its predecessors to urge things forward, grasped at a plethora of initiatives the world might cooperate on with the goal of muting the climate crisis, keeping global temperatures from rising more than 1.5 degrees, and halving CO_2 emissions in the short run on the way to an eventual net zero. Not just countries but destinations and tourism enterprises all took the pledge and became signatories, 850 of them. Their commitments were to:

- Support the global commitment to halve emissions by 2030 and reach Net Zero as soon as possible before 2050.
- Deliver climate action plans within 12 months from becoming a signatory (or updating existing plans), and implement them.
- Align plans with the five pathways of the Declaration (Measure, Decarbonise, Regenerate, Collaborate, Finance) to accelerate and co-ordinate climate action in tourism.

- Report publicly on an annual basis on progress against interim and long-term targets, as well as on actions being taken.
- Work in a collaborative spirit, sharing good practices and solutions, and disseminating information to encourage additional organizations to become signatories and supporting one another to reach targets as quickly as possible.[6]

The World Tourism Organization statement went on to declare:

Everyone in the tourism sector has a role to play in accelerating climate action and therefore all tourism stakeholders (legal entities) can become signatories of the Glasgow Declaration. Destinations (national and local governments), businesses (accommodation providers, tour operators, suppliers, etc.) and supporting organizations (NGOs, business associations, academia, etc.) can become signatories of the Initiative.[7]

Being good can also be good business. If you were one of the signatories of the Glasgow Declaration, you were suddenly able to insert virtuousness into your promotional materials. This, along with a comfortable bed and a spectacular view, could become a selling card to attract travelers wanting to set off on a tourism escape but seeking to do so with a clear conscience. Advice through such vehicles as a feature in Toronto's *Globe and Mail* newspaper on the last day of 2022 followed: "If you want to travel responsibly, here are ten locales doing their best to lessen your impact on the planet."[8] The list of options featured ten destinations in Asia, North and South America, and Europe, each of them a signatory to the World Tourism Organization initiative.

Readers—or potential travelers—might be excused should they catch themselves a tad bemused by the unwitting disingenuousness. Each of the recommended "responsible" trips, it goes without saying, would have to start off with long airplane rides.[9] How else would you get to Finland in order to visit the home of Santa Claus in Rovaniemi, ski or snowboard at some topnotch resorts, or relax in one of the country's three million saunas? Travel by an electric vehicle was recommended upon arrival, "if you're trying to minimize your carbon footprint." Yet, heading to the Brazilian state of Mato Grosso do Sul in order to fish for piranha in the wetlands of Coxim, snorkel in Rio da Prata, or visit one of the biggest flooded caves in the world at Gruto do Lago Azul would also still require the airplane flight. Likewise should one want unashamedly to visit Guanajuato, Mexico, if "looking for history, gastronomy, romance" or Niseko, Japan, if searching out "adventure, gastronomy, nature."[10]

Amid all the clamor about saving the planet, an additional elephant in the room is something rarely acknowledged: there are currently far more of us humans on the face of the earth than ever before, most of us with far greater wealth than anybody in all of history. Coupling this with the explosion of technologies means that almost all of us leave far bigger environmental footprints than any individuals in history. My personal environmental footprint, just in my daily life, is possibly a hundred times that of my ancestors of a mere couple of generations ago—and there are three times as many of my fellow humans as there were when I was born. A heightened sense of the implications of all this has caused us to pause: the impacts of our numbers and our activities on everything from plant and animal species, to the globe's climate, to vulnerable cultures and societies are pushed incessantly in front of us by the media, academia, and simple observation—with a variety of results.

Something I've noticed is an interesting kind of tone creeping into travel writing in the popular press. It suggests a self-consciousness or even an apologetic attitude quite new to the genre. A piece in Canada's *Globe and Mail* had writer Aruna Dutt quoting a peripatetic traveler and photographer, Pat Kane, whose stunning photos from seven recent months of touring through Europe, the Balkans, Fiji, New Zealand, and Southeast Asia were published in the special weekend section of the newspaper. Describing himself as "a self-loathing tourist," Kane declares that he "wonders whether visitors going in droves are ruining the beauty of places like [the ones he has photographed]"—thereby contriving a bit of an ethical pretzel for an inveterate traveler to twist himself into. Kane's way out is to share his excellent photographs and then protest that he will try to go off the beaten path to access "experiences that are more genuine and new . . . not waiting in lines, but getting a better understanding of places and people when you engage with the culture, people and try new food."[11] This sort of rationalization suggests a bit of a fraught future, both ethically and psychologically. Is it going to be possible for us to travel without guilt? Is the future of tourism and travel going to be laden with hypocrisy?

The French adage *plus ça change, plus c'est la même chose* (the more things change, the more they remain the same) might aptly describe a critical feature of the future of travel. The search for the novel will be as much a part of the future as it was of the past. Just as explorers from time immemorial have reached out for shores not previously seen, the quest for the new and "unspoiled" is an abiding mantra for an ever-growing species of today's tourist—as apparently it is for someone like the afore-mentioned photographer. Those who despair that the ancient centers of Rome and Venice along with the cathedrals of England and even the beaches of Hawaii are being turned into theme parks will join the hunt to track down and discover the "unspoiled." This hunt has been ramping up almost hourly to ever more insistent fever pitches. Its urgency seems to come from the thought that this level of imaginative exploration must function as an antidote to some kind of torpor that has set in elsewhere. What many now unabashedly call "adventure tourism" catches its spirit; if you don't want to go along with the herds walking through the Vati-can, you can join an extreme car race to Ulaanbatar, Mongolia. The craze is a winner for its purveyors; the head of marketing for Wild Frontiers, an adventure travel company headquartered in the UK, asserts that they have never been busier.

The Wild Frontiers website declares that the company believes "sus-tainable and immersive travel can be a powerful force for good; breaking down barriers, dispelling myths and bringing people from different cultures and communities together."[12] It enacts this vigorously into the proverbial four corners of the globe, with the result that the number of visitors to Antarctica has more than tripled in the last decade, Nepal has granted a record number of permits to climb Mount Everest, and car rental companies in Mongolia sold out of SUVs.

An accompanying stratagem worth paying attention to is the rapidly growing number of businesses built on support and insurance. Global

Rescue, a company that offers emergency rescues to clients anywhere in the world, reports consumer sales in 2022 to be 36 percent higher than in 2019. Headquartered in Lebanon, New Hampshire, Global Rescue supports clients from operation centers and offices in Manila, Philippines; Islamabad, Pakistan; Amsterdam, Netherlands; Kyiv, Ukraine; and Boston. You get into trouble trying to be cutting edge, they'll save your butt. The principle behind—and the success of—its business is the willingness of travelers to push the envelope and go ever further afield and to ever riskier destinations.

There's a double edge to this sword, of course. Surging interest in adventure tourism is an economic opportunity for a number of formerly out-of-the-way countries that have invested in social media marketing campaigns to bring in visitors. The Mongolian government has invited influencers to come and post videos of the country's verdant valleys, Caribbean-blue lakes, and orange sand dunes. But as we noted a few chapters back with Maya Beach, unspoiled seems to stay unspoiled for only an instant. The embedded irony is that by the very act of bringing in more tourists, the governments endanger their countries' reputations as remote destinations.

Looking to the future, we see an industry whose size is exploding and may well continue to explode exponentially. This has forced many of us, at the very least, into a spot where we need to feel, as we examine our pleasures, that we are being earnestly responsible. We are not about to give up our pleasures, but we believe we need to examine them. If we consider that a conundrum, there is plenty of assistance available to help us through it; myriad enterprises have sprung up to guide and reassure us. A cursory web search will serve up an avalanche of "responsible tourism" services based in all corners of the globe. Culture Trip, a trip booking and guide service, tells us how to "explore the world while being good to it." Culture Trip was launched over ten years ago, in 2011, with a simple yet passionate mission: "to inspire people to go beyond their boundaries and experience what makes a place, its people and its culture special and meaningful. . . . Increasingly we believe the world needs more meaningful, real-life connections between curious travellers keen to explore the world in a more responsible way."

Their statements go on to reassure their audience that they are on the same page in the desire to behave well: "We know that many of you worry about the environmental impact of travel and are looking for ways of expanding horizons in ways that do minimal harm—and may even bring benefits. We are committed to go as far as possible in curating our trips with care for the planet. That is why all of our trips are flightless in destination, fully carbon offset—and we have ambitious plans to be net zero in the very near future."[13] Culture Trip covers all the bases. A list of its "company causes" includes well-being, diversity and inclusion, LGBTQ+, gender equality, sustainability, and family carers.

The cofounder of the UK-based Responsible Travel, Justin Francis, acknowledges that giving his company the name he did back in 2001 was deliberately provocative, "founded on a simple idea: that an industry which accounted for 10 percent of jobs worldwide—that could lift people out of poverty and relied on pristine environments and diverse cultures for its success—had vast potential to be a force for good. Our marketing premise, based on my own travels, was that tourism that benefits local people and places leads to much richer experiences too."[14] Francis is pleased to define his company as "disruptive" and a challenge to the industry. "I don't like much of what I see in tourism, which often tramples over the environment, culture and local people. By exception, our name calls this out and gives customers a choice."[15] He promises, "We screen every holiday against our criteria for responsible travel—travel that maximizes the benefits to local people and reduces negative impacts."

Greenwashing

A wrinkle that goes along with the earnest desires to be correct is the natural human temptation to appear correct even if you're not. Or to, in the interest of public relations, make yourself look better than you really are. In tourism, trying to make your service or facility look more environmentally proper than it is in actuality is called "greenwashing." It's a sizable-enough phenomenon to earn this approbatory designation and everybody—travelers and locals alike—is supposed to be on the watch for it and, when encountered, call it out.

Like the more familiar term "whitewashing," greenwashing conveys a false or misleading impression, in this case about the levels of environmental friendliness of a company's products or services. At its worst, greenwashing goes from simple exaggeration to full-blown misinformation. It is a smudge on the screen of a great deal of serious effort going on to actually make everything from fuels to water use and impacts on forests and beaches less harmful. It is often a scramble among corporate public relations departments—such as those of energy companies—to rebrand, rename, and repackage and hope that the public will be convinced. If they've truly done something to the product, that's great; if it's simply words, then that's "greenwashing."

A FUTURE THAT IS RESPONSIBLE AND AUTHENTIC

A number of organizations are dedicated to research, lobbying, and general public education around the future of tourism and provide additional analysis. Based in Washington, DC, the Center for Responsible Travel (CREST), originally known as the Center for Ecotourism and Sustainable Development (CESD), "provides evidence-based research and analysis to governments, policymakers, tourism businesses, non-profit organizations, and international agencies to solve the most pressing problems confronting tourism, the world's largest service industry."[16] Founded in 2003, the United Nations' International Year of Ecotourism, CREST says it initially focused on the role of small-scale ecotourism in empowering communities and conserving precious resources, but over time evolved to consider how all tourism could be more responsibly planned, developed, and managed across all sectors and geographies. Its vision: "to transform the way the world travels."

There are, as well, a plethora of independent blogs with names like CareElite or Go2Africa that, in turn, give advice or scold their audiences about how to behave or what to keep in mind. The scene is a bit chaotic: Website searches often turn up pages with a mishmash of good and bad, the responsible advisories and the cheap flights websites all turning up shoulder to shoulder. More and more of these enterprises with high purpose are coming out of Africa. Wilderness (once Wilderness Safaris and founded in Botswana in 1983) bills itself as the world's leading conservation and hospitality company, "proud custodians of our planet's most significant wild places."[17]

The British publisher Routledge has devoted substantial effort over a couple of decades to the examination and study of global tourism. In a variety of separate publications, their project has been to define various types of tourism growing in popularity, all of them with a patina of serious responsibility: Ecotourism, Sustainable Tourism, Community-Based Tourism, Fair Trade and Ethical Tourism, Pro-Poor Tourism. In the current vogue, these set up a kind of paradigm within which travelers can self-identify. On one level all this responsibility kind of feels like work, but it is likely to be more and more how matters separate out heading into the future. No more simple sightseeing or kicking back and getting drunk on somebody else's beach.

AUTHENTICITY

A key to the relationship between traveler and host into the future will be not just a heightened sense of responsibility, but also an alertness to "authenticity." Authenticity is one of those nebulous terms. It has to do with what we consider to be "real" (another fraught term). Yet at least some of us think that somehow we can nail it. Its activation is the opposite of hooking into the Metaverse. It is to be out there with minimal defenses and barriers. It seeks more than just a "show." But the question: To what lengths will we go to get real? How far will we push? And what will the reactions of those visited be?

In chapter 8, we wrote about walking tours through the shantytowns of Nairobi. Is this level of more-ardent engagement something new? Is it a good idea? Is it fair to everybody involved—especially the people who are being visited? Many will think it teeters on the edge of what is appropriate, on the edge of voyeuristic, gawking at what the visitor expects to see as human misery. Is it ethical? What is next, tours of war zones? Refugee camps? Yet, in a number of ways, it is merely an extension of people's existing travels for purposes other than tourism, as, for example, members of NGOs (nongovernmental do-gooding organizations) or as journalists. I, under the guise of being a journalist, have been to all three: refugee camps, shantytown slums, and war zones. In those cases my "professional" interest justified, at least in my own mind, my outsider status and even my voyeuristic purposes. The people in the refugee camps had to put up with my asking them questions and taking pictures of them because they too accepted (or were cowed into believing) that I was merely "doing my job."

The do-good tourism where, say, a group of school students from North America go off for two weeks to help some Andean villagers lay the foundations for a school is one step from being a traveler with a professional interest toward what we might call a "semi-professional" interest. Those who crave authenticity will be able to give it a thumbs-up and, one might hope, the visited will accept the engagement in positive terms. The next step, however, to just "go and look," where in the continuum does that sit? Either the continuum of authenticity or the continuum of relationship between visitor and visited? This continues to be a murky area, one that will demand more thought, negotiation, and ethical reflection. If it is a big part of the future of tourism, then it requires some looking at.

The surest prediction one can make about the future of tourism is that it absolutely does have a future. Only the most inveterate curmudgeon among tourism's critics would have the wherewithal to conceive a world in which they, let alone everybody, ceases to travel for pleasure. The act of getting out to see the world and visit other people's cities and countries runs deep in human DNA. If anything, we are destined to travel more, rather than less. What this can only mean is that there is a great deal to take care of.

NOTES

INTRODUCTION

1. Kevin Charach, "David Suzuki Goes on Profanity-Laden Rant," CTV News Vancouver, October 15, 2022, https://bc.ctvnews.ca/david-suzuki-goes-on-profanity-laden-rant-at-federal-government-news-conference-1.6110637.

CHAPTER 1

1. Dean MacCannell, *The Ethics of Sightseeing* (Berkeley: University of California Press, 2011), 63–64.

2. Paul Theroux, *Dark Star Safari: Overland from Cairo to Cape Town* (New York: Houghton Mifflin, 2003), 1.

3. Sallie Tisdale, "Never Let the Locals See Your Map," *Harper's Magazine*, September 1995.

4. Bruce Chatwin, *The Songlines* (London: Franklin Press, 1987), 111.

5. Ibid., 169.

6. Ibid., 171.

7. MacCannell, *Ethics of Sightseeing*, 219.

8. Ibid., 220.

9. Jim Butcher, *The Moralisation of Tourism: Sun, Sand . . . and Saving the World?* (Oxford: Routledge Books, 2003).

10. Publisher's notes for Butcher, *Moralisation of Tourism*, www.routledge.com/The-Moralisation-of-Tourism-Sun-Sand-and-Saving-the-World/Butcher/p/book/9780415296564.

CHAPTER 2

1. U.S. Department of State, U.S. Bilateral Relations Fact Sheet, Background Note: Cuba, May 2007, https://2009-2017.state.gov/outofdate/bgn/cuba/85033.htm.

2. Latin American Network Information Center, Castro Speech Data Base, "Castro Gives Closing Speech at ANAP Congress, May 18, 1992, http://lanic.utexas.edu/project/castro/db/1992/19920518.html.

3. The CUC, introduced in 1994, was eliminated on January 1, 2021, after which tourists and Cubans alike used the moneda nacional, the Cuban peso.

4. The Venceremos Brigades organized for solidarity in 1969 by the Students for a Democratic Society in the United States.

5. Joseph Scarpaci and Armando Portela, *Cuban Landscapes: Heritage, Memory, and Place* (New York: Guilford Press, 2009).

6. Ibid., 116.

7. Florence Babb, *The Tourism Encounter* (Stanford, CA: Stanford University Press, 2011), 4.

8. Larry Krotz, "Wife Shopping," *Saturday Night Magazine*, February 10, 2001.

9. Larry Krotz, "Lovesick," *Saturday Night Magazine*, July 14, 2001.

10. Babb, *Tourism Encounter*, 129.

11. Edward Bruner, *Culture on Tour: Ethnographies of Travel* (Chicago: University of Chicago Press, 2005).

CHAPTER 3

1. Recounted in the *Daily Express*, August 15, 2022.

2. Hugh Honour, *The Companion Guide to Venice* (London: Collins, 1977), 18.

3. Ibid., 19.

4. MOSE Venezia, www.mosevenezia.eu/project.

5. Nikos Zorzos made his case to a number of newspapers including the *Guardian*, the *Telegraph*, the *New York Post*, and the *Irish Times*.

6. Aoife Bradshaw, "Can You Sit on the Spanish Steps? 11 Surprising Ways to Get in Trouble in Italy," Walks, September 12, 2023, www.walksofitaly.com/blog/travel-tips/getting-in-trouble-italy.

7. Eric Reguly, "The Tourist Tidal Wave That's Flattening Europe's Cities, *Globe and Mail*, June 3, 2023.

8. Vicki Brown, "Overtourism in Barcelona," Responsible Travel, www.responsiblevacation.com/copy/overtourism-in-barcelona.

9. Harold Goodwin, "Managing Tourism in Barcelona," Responsible Tourism Partnership Working Paper 1, 3rd ed. (2019), https://responsibletourismpartnership.org/wp-content/uploads/2019/11/Managing-tourism-in-Barcelona.pdf.

10. Elaine McIlwraith, "Desiring 'Convivencia,' Reproducing Overtourism: Sustainability and the Complexity of Tourism Labour in Granada, Spain," paper presented at EASA2020: New Anthropological Horizons in and beyond Europe, July 21–24, 2020.

11. Ibid. Also see Alberto Amore, Martin Falk, and Bailey Ashton Adie, "One Visitor Too Many: Assessing the Degree of Over-tourism in Established European Urban Destinations," *International Journal of Tourism Cities* 6, no. 1: 117–37.

12. Rebecca Aydin, "The History of Airbnb, from Air Mattresses to $31 Billion Company," *Business Insider*, updated September 20, 2019, www.businessinsider.com/how-airbnb-was-founded-a-visual-history-2016-2.

13. Tom Kington," "Venice Cracks Down on AirBnB Tourists as Residents Drift Away from City, *Times* (UK), March 24, 2023.

14. Ibid.

CHAPTER 4

1. "Corporate Information," Carnival Corporation, accessed February 2013, www .carnivalcorp.com/corporate-information.

2. Angela Teberga de Paula and Vania Beatriz Merlotti Herédia, "COVID-19 and Cruise Ships: A Drama Announced," *Études* caribéennes 47 (December 2020), https:// journals.openedition.org/etudescaribeennes/20047?lang=en.

3. "Corporate Information," Carnival Corporation.

4. "New Cruise Ships on Order," Cruise Critic, January 18, 2024, www.cruisecritic.com /articles/new-cruise-ships-on-order.

5. Kim Heacox, "Cruise Ships Are Back and It's a Catastrophe," *Guardian*, July 8, 2021.

6. Nichola Daunton, "Cruise Ships Hurt the Environment, People and Local Communities—and They Don't Pay Taxes, euronews.travel, September 12, 2021, www .euronews.com/travel/2021/12/09/cruise-ships-hurt-the-environment-people-and-local -communities-and-they-don-t-pay-taxes.

7. Friends of the Earth, "Cruise Ships Environmental Impact," March 14, 2022, https: //foe.org/blog/cruise-ships-environmental-impact.

8. Talia Lakritz, "7 Places Being Ruined by Cruise Ships," Business Insider, updated February 5, 2020, www.businessinsider.com/cruise-ships-environmental-impact-tourism -2019-9.

9. Carl Pettit, "Royal Caribbean's Newest Ships to Turn Waste into Energy," Cruise Hive, modified July 11, 2023, www.cruisehive.com/royal-caribbeans-newest-ships-to -turn-waste-into-energy/106110.

10. Bradley Rink, "Liminality At-Sea: Cruises to Nowhere and Their Metaworlds," *Tourism Geographies* 22, no. 3 (July 2019): 392–412.

CHAPTER 5

1. Christopher Wilson, *The Gothic Cathedral: The Architecture of the Great Church, 1130–1530* (London: Thames & Hudson, 1990), 14.

2. The rebuilding of Westminster Abbey in 1245 was paid for entirely by the king, Henry III. This, according to Christopher Wilson, makes it unique among Gothic great churches in Britain, all the rest of which were funded by the (bishopric) corporations that owned them "albeit with the help of donations" (Wilson, *Gothic Cathedral*, 178).

3. Wilson, *Gothic Cathedral*, 7.

4. "The Final Phase of Chichester Cathedral's Major Roof Restoration," Chichester Cathedral, March 6, 2023, https://www.chichestercathedral.org.uk/news/final-phase -chichester-cathedrals-major-roof-restoration.

5. "Chicester Cathedral Receives Grant from Government's Culture Recovery Fund," Chichester Cathedral, October 9, 2020, https://www.chichestercathedral.org.uk/news/ chichester-cathedral-receives-grant-governments-culture-recovery-fund.

6. "Why We Charge for Sightseeing," St. Paul's Cathedral, www.stpauls.co.uk/why-we -charge.

7. Megan Specia, "God Save the Cathedral?" *New York Times*, August 13, 2019.

8. "Norwich Cathedral Helter-Skelter 'Is a Mistake,'" BBC, August 9, 2019, www.bbc .com/news/uk-england-norfolk-49292493.

9. "Rochester Cathedral's Crazy Golf Course Sparks Row," BBC, July 30, 2019, www .bbc.com/news/uk-england-kent-49162116.

10. See the website Thinking Anglicans, www.thinkinganglicans.org.uk.

11. "Visitors Shun Entrance Fee-Charging Cathedrals," Premier Christian News, August 19, 2017, https://premierchristian.news/en/news/article/visitors-shun-entrance -fee-charging-cathedrals.

12. Ecorys produced a study in 2022 for the Association of English Cathedrals that summarized: "What are the economic impacts of cathedrals? Attracting visitors—it is estimated that cathedrals attracted over 9.5 million tourist or leisure visitors in 2019, an increase of 15% on the 2014 total of 8.2 million. For the purposes of the economic impact assessment, this figure excludes worshippers and those taking part in formal educational activity. The additional expenditure generated by these visitors is estimated to be in the order of £128 million in the local economies concerned. Supporting local businesses and economies—in addition, cathedrals are estimated to generate a net contribution of around £107 million in local spending per year (comprising direct, visitor-related and multiplier effects). Creating local jobs—expressing this impact in terms of employment suggests that cathedrals support 5,535 jobs in their local economies. This results in a combined total of approximately £235 million in local spending per year, a slight increase on the £220 million estimated in the 2014 study." (Ecorys, "The Economic and Social Impact of England's Cathedrals," www.englishcathedrals.co.uk/wp-content/uploads /2021/08/Economic-Social-Impacts-of-Englands-Cathedrals-2019.pdf.)

13. Dean MacCannell, *The Ethics of Sightseeing* (Berkeley: University of California Press, 2011), 14.

CHAPTER 6

1. "The Biggest Tourist Traps Worldwide (2023 Data)," Casago, https://casago.com/ blog/biggest-tourist-traps-worldwide.

2. Daniel Macfarlane, *Fixing Niagara Falls* (Vancouver, BC: UBC Press, 2020).

3. Living in Niagara 2020, www.livinginniagarareport.com/living-in-niagara-2020.

4. "Farm Facts about Niagara—Geography," Preservation of Agricultural Lands Society, https://palscanada.org/niagara-fruit-land.

5. Escarpment Fund, "A Brief History of Niagara Region Wine," January 8, 2017, www .escarpmentfund.ca/a-brief-history-of-niagara-region-wine.

6. Ralph Krueger, "The Disappearing Niagara Fruit Belt," *Canadian Geographical Journal*, April 1959.

7. Deborah Reid, "Land Use Changes in a Selected Area of the Niagara Fruit Belt 1954–1978" (master's thesis, McMaster University, September 1980).

8. Living in Niagara 2020, www.livinginniagarareport.com/living-in-niagara-2020.

9. Hugh J. Gayler, ed., *Niagara's Changing Landscape.* (Ottawa: Carleton University Press, 1994).

10. Telephone interview with Gracia Janes, April 2023.

11. Grape Growers of Ontario, Statistics, https://grapegrowersofontario.com/grape -industry-in-ontario/statistics.

12. Louise Elder, PhD, *The History of Canadian Canners Limited, 1903–1986* (Burlington, ON: Canadian Canners Limited, 1986).

13. Email interview with Donald Walker, February 2023.

14. Interview with Debra Marshall, April 2023.

CHAPTER 7

1. "The Value of Tourism in England," Visit Britain, www.visitbritain.org/research -insights/value-tourism-england

2. "About Tourism in Canada," Discover Tourism, www.discovertourism.ca/guidebook /about-tourism-in-canada.

3. OECD Tourism Statistics (database), OECD iLibrary, www.oecd-ilibrary.org/ industry-and-services/data/oecd-tourism-statistics_2b45a380-en.

4. "Kibera Slum Guided Tour from Nairobi," Viator, www.viator.com/tours/Nairobi/ Kibera-Slum-Guided-Tour-from-Nairobi/d5280-15254P27.

5. Dean MacCannell, *The Ethics of Sightseeing* (Berkeley: University of California Press, 2011), 66.

6. "Tourism in France," Ministère de l'Europe et des Affaires Étrangères, www .diplomatie.gouv.fr/en/french-foreign-policy/tourism/tourism-in-france.

7. Martin Mowforth and Ian Munt, *Tourism and Sustainability: Development, Globalisation and New Tourism in the Third World* (London: Routledge, 2016), 1.

8. Karla Cripps, "Tourism Killed Thailand's Most Famous Bay. Here's How It Was Brought Back to Life," CNN, August 1, 2022, www.cnn.com/travel/article/maya-bay -thailand-recovery-c2e-spc-intl/index.html.

9. World Travel & Tourism Council, https://wttc.org.

CHAPTER 8

1. Karen Blixen, aka Isak Dinesen, from *Out of Africa*, 1937.

2. See, for example, www.ecolyx.com, www.eastafricasafariventures.com, and https:// thenextcrossing.com.

3. Steve King, "Getting into the Stride: Walking in Kenya's Loita Hills," *Condé Nast Traveller*, February 13, 2021, www.cntraveller.com/article/loita-hills-kenya-walking.

4. "Loita Hills / Maasai Trails," New African Territories, www.africanterritories.co.ke /maasai-trails.

5. "Day 1–2 Arusha—Maasai Village, 'Olpopongi,'" Safari Dynamics, https://safari -dynamics.wixsite.com/safari-dynamics/cultural-tours.

6. Alexis Bunten and Graburn, Nelson, eds., *Indigenous Tourism Movements* (Toronto: University of Toronto Press, 2018), 12.

7. Ibid., 59.

CHAPTER 9

1. "About Us," Stoney Nakota Nations, https://stoneynakodanations.com/about-us.

2. Jonathan Clapperton teaches Canadian studies in the College of Interdisciplinary Studies at Royal Roads University in Victoria, BC, and is an adjunct professor at the

University of Victoria. He was quoted in "Banff Indian Days Affirmed Stereotypes, Reinforced Culture," *Rocky Mountain Outlook*, April 3, 2014, www.rmoutlook.com/local-news/banff-indian-days-affirmed-stereotypes-reinforced-culture-1565423.

3. Ibid.

4. Alexis Bunten and Nelson Graburn, eds., *Indigenous Tourism Movements* (Toronto: University of Toronto Press, 2018), 3.

5. Secretariat of the Permanent Forum on Indigenous Issues, "The Concept of Indigenous Peoples," background paper presented at the United Nations Workshop on Data Collection and Disaggregation for Indigenous Peoples, January 12–21, 2004, https://www.un.org/workshop_data_background.

6. Bunten and Graburn, *Indigenous Tourism Movements*, 10.

7. Ibid., 21.

8. Ibid., 12.

9. Estelli M. Smith, "Tourism and Native Americans," Cultural Survival, February, 9, 2010, www.culturalsurvival.org/publications/cultural-survival-quarterly/tourism-and-native-americans-63.

10. Cristell Bacilio, "Hawaii Tourism: Opposite of a Paradise for Locals," International Relations Review, October 5, 2022, www.irreview.org/articles/hawaii-tourism-opposite-of-a-paradise-for-locals.

11. Ibid.

12. See "Banff Indian Days," *Rocky Mountain Outlook*, April 3, 2014.

13. "Daily Cultural Experiences," Wikwemikong Tourism, https://wikytours.com/daily-cultural-experiences.

14. Telephone interview with Kevin Eshkawkogan, August 14, 2023.

15. "Who We Are," Indigenous Tourism Ontario, https://indigenoustourismontario.ca/who-we-are.

CHAPTER 10

1. Susan Sontag, *On Photography* (New York: Farrar Straus and Giroux, 1977), 127.

2. Ibid., 18.

CHAPTER 11

1. *Condé* Nast Traveler, 35th anniversary special edition, *The* Future of Travel, September/October 2022.

2. Toby Skinner, "What Is the Future of Travel in the Metaverse?" *Condé* Nast Traveler, September 7, 2022, www.cntraveler.com/story/what-is-the-future-of-travel-in-the-metaverse.

3. This was written before Branson's Virgin Orbit Holding filed for bankruptcy in April of 2023.

4. Mary Holland, "These 7 Hotels Are Pioneering More Responsible Travel, *Condé Nast Traveler*, September 7, 2022, www.cntraveler.com/story/7-hotels-pioneering-more-responsible-travel.

5. Elissa Garay, "These 5 Airlines Are Using the Latest Technology to Make Their Planes Greener," *Condé* Nast Traveler, September 7, 2022, www.cntraveler.com/story/these-5-airlines-are-using-the-latest-technology-to-make-their-planes-greener.

6. UN Tourism, "The Glasgow Declaration on Climate Action in Tourism," www.unwto.org/the-glasgow-declaration-on-climate-action-in-tourism.

7. Ibid.

8. Jennifer Foden, "Where to Travel in 2023 if You Care about Climate Change," *Globe* and Mail, December 30, 2022, www.theglobeandmail.com/life/travel/article-travel-sustainability-climate-change.

9. According to the David Suzuki Foundation, a quarter of all greenhouse emissions by 2050 could be from airplanes. While many sectors are reducing their emissions, those from aviation have continued to grow. Carbon emissions from the airline industry grew by 75 percent from 1990 to 2012. This is expected to be the case until 2050.

10. Foden, "Where to Travel in 2023."

11. Aruna Dutt, "Around the World in Eight Photographs," *Globe* and Mail, July 86, 2023, www.theglobeandmail.com/life/travel/article-travel-photography-pat-kane.

12. Wild Frontiers, "Our Purpose," www.wildfrontierstravel.com/en_US/our-purpose.

13. Culture Trip, "Diversity, Equity & Inclusion at Culture Trip," https://theculturetrip.com/europe/articles/diversity-equity-inclusion-at-culture-trip.

14. Justin Francis, "Responsible Travel: How Sustainable Tourism Has Changed Over the Past 20 Years," *Independent* (US), April 6, 2021, https://www.independent.co.uk/climate-change/sustainable-living/responsible-travel-anniversary-sustainable-tourism-b1827423.html.

15. Responsible Travel, www.responsibletravel.com.

16. Center for Responsible Travel, www.responsibletravel.org.

17. Wilderness, www.wildernessdestinations.com.

BIBLIOGRAPHY

Adams, Kathleen. *The Ethnography of Tourism: Edward Bruner and Beyond*. Lanham, MD: Lexington Books, 2019.

Babb, Florence. *The Tourism Encounter*. Stanford, CA: Stanford University Press, 2011.

Blixen, Karen (Isak Dinesen). *Out of Africa*. Copenhagen: Gyldendal and London: Putnam, 1937.

Bruner, Edward. *Culture on Tour: Ethnographies of Travel*. Chicago: University of Chicago Press, 2005.

Bryson, Bill. *The Road to Little Dribbing: More Notes from a Small Island*. New York: Doubleday, 2015.

Bunten, Alexis, and Nelson Graburn, eds. *Indigenous Tourism Movements*. Toronto: University of Toronto Press, 2018.

Butcher, Jim. *The Moralisation of Tourism: Sun, Sand . . . and Saving the World?* Oxford: Routledge Books, 2003.

Cameron, Sarah. *Explore Cuba*, Insight Guides. London: APA Publications, 2018.

Chatwin, Bruce. *The Songlines*. New York: Viking, 1987.

de Botton, Alain. *The Art of Travel*. London: Penguin, 2002.

Fermor, Patrick Leigh. *A Time of Gifts*. London: John Murray, 1977.

Figueredo, D. H. *The History of the Caribbean*. New York: Facts on File, 2008.

Gayler, Hugh J., ed. *Niagara's Changing Landscape*. Ottawa: Carleton University Press, 1994.

Honour, Hugh. *The Companion Guide to Venice*. London: Collins, 1977.

Kingsley, Mary. *Travels in West Africa*. London: Penguin Classics, 2015 (originally published in 1897).

MacCannell, Dean. *The Ethics of Sightseeing*. Berkeley: University of California Press, 2011.

———. *The Tourist: A New Theory of the Leisure Class*. New York: Schocken Books, 1989.

Macfarlane, Daniel. *Fixing Niagara Falls*. Vancouver, BC: UBC Press, 2020.

Meinhardt, Maren. *Alexander von Humboldt: How the Most Famous Scientist of the Romantic Age Found the Soul of Nature*. New York: BlueBridge, 2019.

Mowforth, Martin, and Ian Munt. *Tourism and Sustainability: Development, Globalisation and New Tourism in the Third World*. London: Routledge, 2016.

Parr, Martin. *Small World*. Stockport, UK: Dewi Lewis Publishing, 2007.

Pearen, Shelley J. *Exploring Manitoulin*. 3rd ed. Toronto: University of Toronto Press, 2001.

Scarpaci, Joseph, and Armando Portela. *Cuban Landscapes: Heritage, Memory, and Place.* New York: Guilford Press, 2009.

Scherf, Kathleen, ed. *Adventures in Small Tourism.* Calgary: University of Calgary Press, 2023.

Sontag, Susan. *On Photography.* New York: Farrar Straus and Giroux, 1977.

Stemp, Richard. *The Secret Language of Churches and Cathedrals.* London: Duncan Baird Publishers, 2010.

Theroux, Paul. *Dark Star Safari: Overland from Cairo to Cape Town.* New York: Houghton Mifflin, 2003.

Wilson, Christopher. *The Gothic Cathedral: The Architecture of the Great Church, 1130–1530.* London: Thames & Hudson, 1990.

PAPERS

Goodwin, Dr. Harold. "Managing Tourism in Barcelona." Responsible Tourism Partnership Working Paper 1, 3rd ed. (2019). https://responsibletourismpartnership.org/wp-content/uploads/2019/11/Managing-tourism-in-Barcelona.pdf.

McIlwraith, Elaine. "Desiring 'Convivencia,' Reproducing Overtourism: Sustainability and the Complexity of Tourism Labour in Granada, Spain." Paper presented at EASA2020: New Anthropological Horizons in and beyond Europe, July 21–24, 2020.

Secretariat of the Permanent Forum on Indigenous Issues. "The Concept of Indigenous Peoples." Background paper presented at the United Nations Workshop on Data Collection and Disaggregation for Indigenous Peoples, January 12–21, 2004. https://www.un.org/workshop_data_background.

Index

Airbnb, 54, 58–61, limits on 62, 105
Alhambra palace and fortress, 56
Anishinaabe people, 135
Artania, 65–66
Ashenden, Dr. Gavin, 86
Australian government, 66

Babb, Florence, 36, 39
Bacilio, Cristell, 133
Ballano, Ada Colau, 55
Banff Indian Days, 128
Banff National Park, 127, 134
Barcelona, 55
The Beach, 20, 112
Bethlehem, Church of the Nativity, 89
Billy family, 11
Boorstin, Daniel, 22
Boulter, Wellington, 99
British Columbia, 4
Brown, Vicki, 55
Brugnaro, Luigi, 61
Bruner, Edward, 41
Bryant, Rev Canon Andy, 85
Bunten, Alexis, 123, 131–132

Burton, Richard (Anatomy of Melancholy), 19
Butcher, Jim, 22

Canada, tourism in GDP, 106
Canadian Pacific Railway (CPR), 127
Canterbury Cathedral, 82
Carnival Cruise Lines, 64, 66, 71
Castro, Fidel, 26, 32, 39
Castro, Raul, 32
Center for Responsible Travel, 160
Chatwin, Bruce, 13, 18–19
Chester Cathedral LEGO project, 85
Chichester Cathedral, 79, 82–83
Church of England cathedrals, 7
Cistercian monks, 81
Clapperton, Jonathan, 128
Club Med, 10
CNN, 1, 42
Conde Nast Traveler, 151
COVID pandemic, 1, 42, 64–65, 73, 87
Cruise Lines International Association, 153
Cruise Mapper, 66

Cruise Mummy, 68, 75
Cuba, 1, 5, 25; economy 27; tourism 28; Russians in Cuba 28; mafia in 29; all-inclusive resorts 30, 36
Cuba Tourist Board, 30
Cuban tourism economic numbers, 31
Culture Trip, 158–159

Darwin, Charles, 14
Daunton, Nichola, 69
Debassige, Neil, 136
Declaration of the Rights of Indigenous People, 131
Diamond Princess cruise ship, 65
Díaz-Canal, Miquel, 32
DiCaprio, Leonardo, 20, 112
Disney Cruises, 72, 75
Dominican Republic, 39
Dubrovnik, 143
Durham Cathedral, 7, 79, 84
Dutt, Aruna, 156

Emerson, Ralph Waldo, 19
Eshkawkogan, Kevin, 141–142

Facebook, 12, 149
Fermor, Patrick Leigh, 13–14
Fodor, 3
France, tourism in GDP, 106, 110
Francesco 1, 67
Francis, Justin, 159
Friends of the Earth, 70

Glasgow Declaration on Climate Action in Tourism, 154–155
Global Rescue, 158
Gloucester Cathedral, 7, 83–84
Gomez Molina, Lessner, 31
Goodwin, Dr. Harold, 56
Graburn, Nelson, 123, 131–132
Granada, 57–58
Greece, tourism in GDP, 106
Greenwashing, 159
Guevara, Che, 34

Havana, 36–38
Hawaii, 133
Hawaiians (native), 9
Heacox, Kim, 69
Hemingway, Ernest, 34
Hereford U.K., 77
Hereford Cathedral, 77–78
Holland America Line, 75
Honour, Hugh, 47

Indigenous, definition, 130–131
Indigenous Tourism Ontario (ITO) 141
Instagram, 149
Israel, tourism in GDP, 106

Janes, Gracia, 97
Jibacoa, 27–28

Kane, Pat, 156
Kapuscinski, Ryszard, 13
Kierkegaard, Soren, 19
King, Steve, 120–121

Kingsley, Mary, 13–14
Korea, 15
Kwakiutl, 11

leakage, 114
Lonely Planet guidebooks, 3

Maasai, 9–10, 117–125, 132
Maasai Trails Safari
    Company, 121
MacCannell, Dean, 12, 21, 89, 109
Macfarlane, Daniel, 93
Manitoulin Island, 134
Mappa Mundi, 77
Marghera, 49
Marshall, Debra, 100–101
Masai Mara, Loita Forest, 118
Maya Bay, Thailand, 20, 112
McCarthy, Mary, 49
M'Chigeeng First Nation,
    136, 141
McIlwraith, Elaine, 57–58
Metaverse, 12–13
Minchin, Louise, BBC, 87
Moors, 57
MOSE (Modulo Sperimentale
    Elettromeccanico), 51
Mowforth, Martin, 111
MSC Cruises, 72
Munt, Ian, 111

Nairobi, shantytown tourism,
    108–109
New York State, 8
Niagara Escarpment, 8, 94

Niagara Falls, 8, 92
Norfolk Cathedral, helter skel-
    ter, 85
North American Free Trade
    Agreement (NAFTA), 8, 96
Norwegian Cruise Lines, 75

Ojibway Cultural Foundation,
    139, 142
On Photography, 147
Ontario, 9
Opawamick, Hope, 139
over-tourism, 53
Oxford, England, 15
Oxford English Dictionary, 16

P & O cruise line, 67
Paris, 6
Phillips, Rev. Rachel, 86
Picton, 99
Portela, Armando, 32
Preservation of Agricultural Lands
    Society, 94, 97
Prince Edward County, 98

Recollet, Naomi, 139–140
Reguly, Eric, 54
Rink, Bradley, 76
Rochester Cathedral, mini golf, 85
Rollinmud, Roland, 134
Roman Catholic cathedrals, 7
Rome, 6, 52, Spanish Steps, 52
Routledge Publishing, 113, 161
Royal Caribbean Cruise Line, 65,
    68, 71, 75

Safari Dynamics, 122
Said, Edward, 40
Salazar, Noel B., 124
Sangwine, Rev Geoffrey, 88
Santorini, 51
Sasson, Steven, 144
Scarpaci, Joseph, 32
sex tourism, 38–39
Skinner, Toby, 151–152
Smith, M. Estellie, 132–133
Sontag, Susan, 147
Specia, Megan, 85
Steves, Rick, 13
Stoney Nakoda First Nation,
    127–129, 134
St Paul's Cathedral, 84–85
sustainable aviation fuel, 153
Suzuki, David, 5

Thamrongnawasawat,
    Dr. Thon, 112
Thaver, Sherali, 30
Theroux, Paul, 13, 15
Thomas Cook, 10, 22 rail excur-
    sions, 68 cruises
Tisdale, Sallie, 16

Tourismteacher.com, 4, 123

U.K. tourism in GDP, 106

Vancouver, 4
Varadero, 30
Venice, 2, 6, 45, cruise ships 48,
von Humboldt, Alexander, 14, in
    Cuba 28

Walker, Donald, 100–101
Welles Cathedral, 82
Westminster Abbey, 7, 79, 87
Wikwemikong (Wiikwemkoong),
    135–138
Wild Frontiers, 157
Wilson, Christopher, 81–82
World Bank, Yearbook of Tourism
    Statistics, 3–4
World Travel &Tourism
    Council, 113
World Tourism Organization
    (UNWTO), 2, 155

Zorzos, Nikos, 52
Zuckerberg, Mark, 12

About The Author

Larry Krotz is a journalist, documentary maker, and author of more than a dozen books on topical issues. He has written for a number of magazines as well as the *Globe and Mail* newspaper and made documentaries that were aired on PBS and distributed by the National Film Board of Canada. Throughout his career, he has brought a perceptive and careful eye to what matters to our culture. His first book on tourism in 1996 identified, in the words of one reviewer, how "the current demand for inconvenience-free, responsibility-free travel is threatening the very sights people spend billions of dollars a year to visit." His 2008 book, *The Uncertain Business of Doing Good: Outsiders in Africa*, showed what happens when good intentions do not always have the best outcomes. Recent books in 2015 and 2021 traced the stories of type 2 diabetes in Indigenous children in the far north (*Diagnosing the Legacy*) and the creation of a specialized medical school in northern Ontario (*Nothing Ordinary*). Larry Krotz lives in Toronto, Canada. His website is www.larrykrotz.ca.